AMAZINGLY AMUSING A-Z ANIMAL FACTS FOR KIDS (AND THEIR ADULTS!)

WILDLIFE BIOLOGY WITH A HUMOROUS TWIST!

THE BONKERS BIOLOGIST

© **Copyright the Bonkers Biologist 2021 - All rights reserved.**

The content contained within this book may not be reproduced, duplicated or transmitted without direct written permission from the author or the publisher.

Under no circumstances will any blame or legal responsibility be held against the publisher, or author, for any damages, reparation, or monetary loss due to the information contained within this book. Either directly or indirectly. You are responsible for your own choices, actions, and results.

Legal Notice:

This book is copyright protected. This book is only for personal use. You cannot amend, distribute, sell, use, quote or paraphrase any part, or the content within this book, without the consent of the author or publisher.

Disclaimer Notice:

Please note the information contained within this document is for educational and entertainment purposes only. All effort has been executed to present accurate, up to date, and reliable, complete information. No warranties of any kind are declared or implied. Readers acknowledge that the author is not engaging in the rendering of legal, financial, medical or professional advice. The content within this book has been derived from various sources. Please consult a licensed professional before attempting any techniques outlined in this book.

By reading this document, the reader agrees that under no circumstances is the author responsible for any losses, direct or indirect, which are incurred as a result of the use of the information contained within this document, including, but not limited to, errors, omissions, or inaccuracies.

© theBonkersBiologist.com – All rights reserved. Do not share, copy, or sell any part of this document unless you have written permission from theBonkersBiologist.com. All infringements will be prosecuted.

*For Rosie, Jimbo, and Clem... you're where it all started.
For Tassie... you followed your dreams and encouraged me to follow mine.
For the Ditamores and Carters... you embraced me as your own.
For Steev and Jack... you changed my life.*

TABLE OF CONTENTS

Introduction 1

    Chapter A: Armadillo and Arapaima 3

    Chapter B: Bee 9

    Chapter C: Cockroach 15

    Chapter D: Dolphin 21

    Chapter E: Electric Eel and Echidna 29

    Chapter F: Frog 37

    Chapter G: Galápagos Tortoise and Greenland Shark 41

    Chapter H: Hydra and Horseshoe Crab 47

    Chapter I: Indian Giant Squirrel and Indian Palm Squirrel 55

    Chapter J: Jerboa 61

    Chapter K: Kookaburra 65

    Chapter L: Loggerhead Shrike and Lobster 69

    Chapter M: Mantis Shrimp 77

    Chapter N: Nudibranch 81

    Chapter O: Opossum and Oyster 85

    Chapter P: Platypus and Paddlefish 93

    Chapter Q: Queensland Grouper 99

    Chapter R: Ruby-Throated Hummingbird 103

Chapter S: Sea Turtle 109

Chapter T: Termite and Tamandua... 115

Chapter U: Uakari and Unau 123

Chapter V: Vampire Squid and Virgin Islands Dwarf Gecko... 131

Chapter W: Whelk and Whale Shark 139

Chapter X: Xingu River Ray and Xanthippe's Shrew 145

Chapter Y: Yeti Crab.. 151

Chapter Z: Zebra Shark and Zigzag Salamander 157

In Closing... 163

References... 165

A FREE GIFT FOR YOU!

Amazingly Amusing Animal Trivia for Kids (and Their Adults!)

Test your knowledge of all the amazingly amusing facts you'll learn in this book! Quiz questions for older kids (and parents!), coloring pages for younger kids!

Scan the QR code or visit this link:

theBonkersBiologist.com/animal-trivia

INTRODUCTION

There are over 1,371,000 animal species on Earth. I know this because I'm a wildlife biologist (and I know how to use Google!). I'm also a mom... a reluctant mom, I'll admit, because at first it was hard to set aside work to focus my attention on keeping a brand new human alive and thriving.

But, man, was the sacrifice worth it! I have loved introducing my kiddo to the natural world around him. We love to hike nature trails, go beach combing, watch lizards in our yard, catch and release bugs that get inside the house... anything and everything involving animals!

Over the course of my career, I implemented conservation projects for birds and marine animals, fought to pass federal natural resource legislation, and distributed millions in grant funding for fish and wildlife conservation programs worldwide.

Throughout my years and years of working with animals and implementing wildlife conservation projects, I had the privilege to encounter some fascinating creatures in person and observe their behaviors up close. I couldn't keep all those stories to myself, so I wrote them down for others to enjoy, too.

"If we can teach people about wildlife, they will be touched. Share my wildlife with me. Because humans want to save things that they love."

~STEVE IRWIN

THE BONKERS BIOLOGIST

I'm stoked to share all these amazingly amusing animal facts with you! My goal is to inspire a love and respect for this planet that will make the next generation socially and environmentally conscious and motivate them to protect the earth, its inhabitants, and its resources. Hopefully, you'll also get a good laugh and learn some new things in the process!

Let's gooooooooo!

CHAPTER A
Armadillo and Arapaima

Armadillo
(aar·muh·**di**·low)

A rmadillos are funny looking creatures! With long snouts, armored plates covering their bodies, and the tail of a rat, who wouldn't find that amusing?

Armadillos are mammals, animals that are characterized by body hair or fur. But armadillos have hard shell-like scutes protruding from their skin, so what's the deal? Well, those scutes are bony plates called osteoderms, and they're coated in a layer of keratin. What's keratin, you ask? It's the protein that makes hair!

If you look closely, you can see individual hairs sprouting out from between these plates, and there are especially thick patches of hair on the underside of an armadillo's body, where there aren't any scutes. Without armor on its underside, the armadillo's fleshy belly isn't as well protected as its head, back, and butt. So, if it encounters a predator, it will hunch close to the ground to protect its soft underside while relying on the armor to protect its topside. One species, the three-banded armadillo, is flexible enough that it can actually roll into a ball, folding its scutes around its entire body to create full-body protection!

Let's move on to nine-banded armadillos. Aside from just being funny looking in general, the most amusing thing about them is that females always give birth to identical quadruplets! The odds of a human having identical quads are one in 15 million, but, barring any problems in the womb, these armadillos do it for each and every pregnancy.

Another cool thing about female armadillos is that they "save" fertilized eggs for a few months before actually becoming pregnant. So, let's say a pair mates in early winter. It won't be until early spring before mama releases her fertilized eggs into the womb so they can implant and begin developing over a period of approximately five months. The total time from mating to birth is 9-ish months, just like humans.

But what if springtime brings a tough season full of drought and food scarcity? Those aren't the best conditions to nourish a pregnant armadillo and her developing quads. So, what's a mama with a killer maternal instinct to do? She delays the implantation of embryos even longer than the typical 3-4 months – sometimes up to 2 years! Armadillos are the only mammal in the world known to have this ability.

Instead of implanting in the uterine wall, the tiny embryo just floats around in a dormant state until all the environmental stressors are gone. Then, once the mama armadillo's stress hormones normalize, the embryo implants and begins developing as if nothing happened. So, essentially, an armadillo can be pregnant for years before giving birth! Two years of being pregnant and then birthing four babies? No, thank you!

Back to an armadillo's armor… you may have heard that it's so strong, it's bulletproof. Sorry to burst your bubble, but that's an old wives' tale. That said, armadillos *have* inspired scientific innovations in the development of human body armor. Pretty cool, huh?

One final weird – and random – fact about armadillos? They are one of only two animal species in the world that can contract leprosy, the other being humans!

Arapaima
(eh·ruh·**pai**·muh)

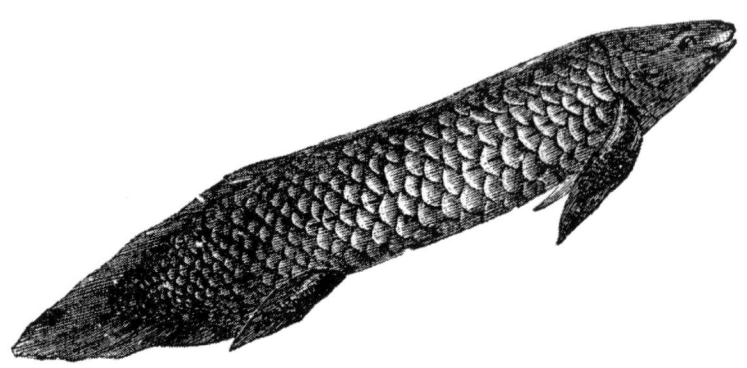

Arapaimas are thought to be the largest freshwater fish in the world. They're found in several countries in northern and central South America and can weigh many hundreds of pounds. According to the International Game Fish Association (IGFA), which keeps official world records of fish catches, the largest arapaima caught by an angler was a 339.5 pounder back in 2010!

There are at least four species of arapaima. I say "at least" because the most recent species was just discovered in 2013. Who knows if it'll be the last?! The largest of the arapaima species, *Arapaima gigas*, is hunted for food in the Amazon. Locals eat their scales, meat, and even their tongues! The tongue thing is especially weird because arapaima tongues have a large bone in them. It's used to pulverize their prey against teeth in the roofs of their mouths before swallowing!

The way they catch food is weird, too. Instead of chasing and then grabbing prey in their teeth, arapaimas suck their prey into their mouths in large gulps. Typically, they'll corner a smaller fish up at the water's surface, or find animals like insects on *top* of the water's surface, then take a large gulp of water, sucking the animal right in!

Arapaima scales are so hard that they're more like bony plates. As such, they don't have many predators. In fact, the caiman – a member of the alligator family – is the only animal known to feast on them… except for humans. In the mid-2000s, arapaima began being farmed for export. They can now be found on the menus at specialty restaurants in the U.S. and elsewhere, and they were even featured as the "secret ingredient" on the Food Network's Iron Chef America cooking show in 2011!

Another unreal fact about arapaimas? They breathe air through their mouths! Baby arapaimas have gills, but they become less pronounced as the fish ages before finally disappearing altogether. At that point, arapaimas have to come to the surface for air. They can hold their breath up to 20 minutes, if need be, but prefer to stay around the surface of the water so they can breathe as often as they like. Because they're mouth-breathers, they can also live out of water for a long time – up to a full day!

While most fish don't stick around after their eggs hatch, arapaimas do. And not just mom… male arapaimas take an active role in caring for their hatchlings, also called "fry." Most notably, they hold their babies inside their mouths to protect them from predators!

THE BONKERS BIOLOGIST

Although mature arapaimas don't have many predators, their young do. Plenty of fish would be happy to make a meal of them. So, dad not only keeps them close, but he'll also gobble them up to hide them if a predator is nearby. His body even changes color to match the dark hue of his babies, further camouflaging them! When the danger has passed, dad releases the little fry to swim back out into the water. That kind of paternal devotion is incredibly rare in fish!

CHAPTER B
Bee
(bee)

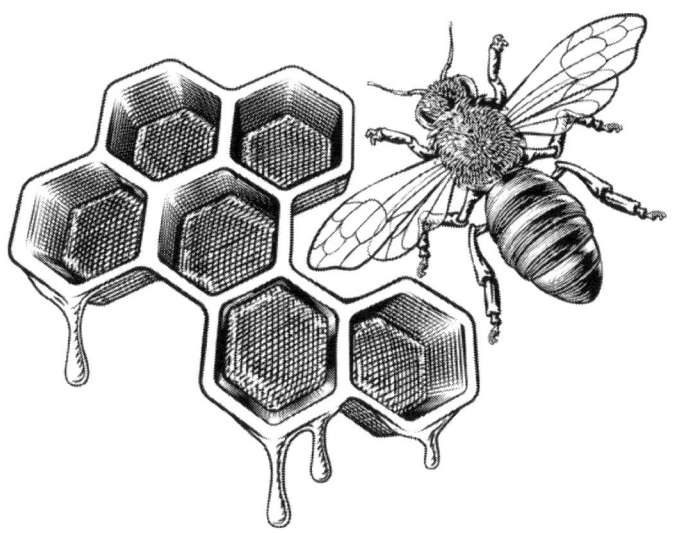

Do you like honey? Ah, honey... sweet, golden nectar of the Gods... er, I mean, of the bees. But how do they make it? Buckle up; this is gonna be good!

You may already know that bees are Earth's #1 pollinators. They move from flower to flower, collecting pollen and nectar. Pollen is a primary source of protein for bees, and nectar is their primary source of carbohydrates. The bees aren't actively trying to pollinate; they just end up transferring pollen to different flowers as they go, which helps fertilize the next generation of plants and crops. The pollination they provide – a sort of "side effect" of the bees' pollen and nectar collection – is something that humans depend on for survival!

There are over 4,000 species of bees, and they aid in pollination by mixing the pollen between the male and female parts of flowers, which results in the creation of seeds. Honey bees are the main heroes here – they pollinate around 35% of all crops and 80% of all flowering plants on Earth. Most of our food comes from flowering plants, so humans literally depend on pollinators for food. Bees account for $15 billion of annual crop production in the U.S. – that's billion with a B! And that doesn't count the value of the honey they produce annually – which comes in at a cool $300 million plus per year!

But back to honey. Honey bees have one of two roles – they're either forager bees or house bees (also called drone bees or worker bees). Forager bees leave the hive to find nectar, which they suck up from flowers via a long, straw-like tongue called a proboscis. They store the nectar in their stomachs as they collect it, visiting 50-100 flowers before returning to the hive.

Foragers take about a dozen trips a day to collect pollen and nectar, visiting well over a thousand flowers and carrying back a total daily load of pollen that's four times their body weight! A single colony can collect over 125 pounds of pollen a year. Not only that, but foragers travel looooong distances… two to five miles each trip. At a dozen trips a day, that's 24-60 miles their tiny wings fly each and every day!

Nectar is basically sugar water, a very fluid, easily pourable solution. But honey is thick and sticky. So, how does that process happen?

As the forager bees are working to collect nectar, their bodies are working to make it into honey. As bees suck up nectar, an enzyme in their salivary glands is added that begins the process of converting sucrose, the complex sugar found in nectar, into simple sugars called glucose and fructose. This process continues in the bees' stomachs.

Once back at the hive, they regurgitate the processed nectar for the house bees. The house bees then take over the processing, adding more enzymes as they swallow and regurgitate the solution over and over again. They also remove excess water from the nectar to concentrate it, making it into the sticky, syrupy, golden honey that we're familiar with.

Then, the house bees deposit it into honeycomb tunnels, double-check that all the moisture has evaporated, and seal the comb with wax. Bees secrete wax from glands in their tummies.

You may have noticed I mentioned regurgitation a couple of times now... you know that's a fancy word for vomit, right? Yep, when you eat honey, you're eating *bee vomit*!

In 2020, there were about 2.9 million bee colonies in the U.S., which collectively produced over 150 million pounds of honey. This is a typical annual yield, which is pretty impressive considering that a single bee only makes about 1/8 teaspoon of honey over its lifetime. Enjoy that statistic the next time you smother your toast with delicious bee barf!

Before moving on, there's one more amazing bee fact I want to share. I recently had a colony of honey bees removed from my house – they'd found a hole behind the outer siding of the house and managed to get into the joists between the first and second floor. I called a beekeeper as soon as possible and, less than 48 hours after I saw the first sign of the colony, she arrived to remove and relocate them.

When our beekeeper cut out a hole in the ceiling to remove them, we found an impressive little hive! There were between 7,000 and 8,000 bees hanging off of 3 combs they'd built in the short time they'd been there. She scooped them out – by hand and with no head net! – into a bee box she'd brought along that had some honey in it from the beehives she tended at her house. Since they'd only arrived less than two days prior, the bees between my floor joists hadn't produced honey yet – they were still busy making the comb. So, they seemed happy to move into a bee box that already had honey in it.

After scooping out as many bees as possible, the keeper then cut their combs from the joists and inserted them into frames inside the bee box. Any remaining bees just followed their combs right into the box. She did all this without any protection and without getting a single sting!

We left the box by the ceiling hole until the evening, just to make sure any bees that had been out foraging during the hive removal had somewhere to return to. Once the sun set, the beekeeper double-checked that no bees remained in the ceiling, closed up the box, and took the bees to their new home in one of her apiaries.

Naturally, I asked her a thousand questions while she was at the house. One crazy thing I learned (one of many) is that colonies have occasionally been known to engage in mutiny! I always assumed the queen bee was just that – the queen bee – and that that didn't change until she groomed a new queen to take her place when she died. But I was wrong!

If a colony doesn't like their queen – if she's old or isn't as productive as they'd like or otherwise isn't up-to-snuff – worker bees can go behind her back to rear a new queen to overtake her. When she isn't watching, they'll remove an egg from a random honeycomb cell and place it into one of the queen cells. After it hatches, it will force the undesirable queen out (S. Daye, personal communication, July 25, 2021)! Simply ruthless, right? Bees are a trip!

CHAPTER C
Cockroach
(**kaa**·krowch)

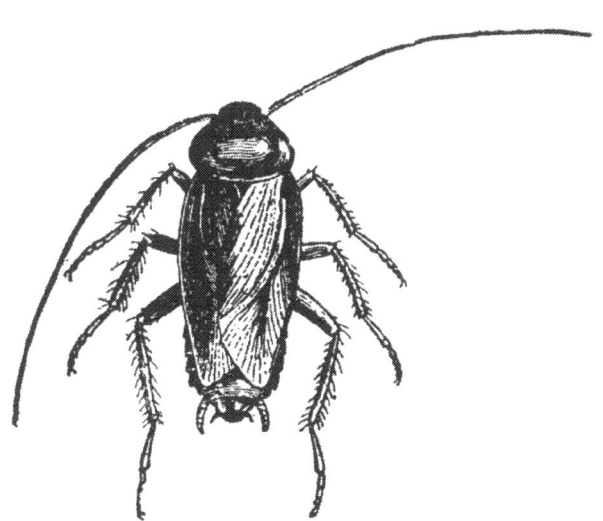

I bet I can guess what you're thinking… "Why on Earth would she feature *cockroaches*, of all things?!" Well, as gross as this wildlife biologist thinks they are, and as much as I don't want them in my home or anywhere near me, they're still fascinating creatures that have mastered the whole "survival of the fittest" thing.

Cockroaches have been around for about 300 million years – before dinosaurs! Thousands of different species have evolved into existence since then, and there are over 4,500 species of cockroaches on the planet today. They're super adaptable and can be found living all over the world, from the hottest climates at the equator to the coldest environs of the arctic!

Have you heard the old wives' tale that cockroaches can survive nuclear bombs? Well, it's partially true, though not entirely accurate. If a cockroach were to be at ground zero of a blast, it would 100% be fried like the rest of us. But they *can* handle the long-term radiation from a nuclear fallout about ten times better than humans. Meaning that it takes ten times the level of radiation to kill a cockroach than it does to kill a human.

I mean, these guys are tough. They can go a month without eating, about ten days without water, and up to ten days without their heads.

Wait, what?! You read that right… cockroaches don't need their heads to live! At least, not for a while. They breathe through small holes in their bodies that are connected to their tracheas, so they don't need their heads to breathe, only to eat and drink. If it weren't for dying of thirst, a headless cockroach might even be able to live longer than ten days!

Some cockroach species have been clocked running at speeds of 5 feet per second. That's 300 feet per minute, which is 3.4 miles per hour! Incredibly fast for an insect. Once they really get going, they sometimes rise up onto their two hind legs, like humans. Taking into account and adjusting for our size differences, a cockroach running 3.4 miles per hour is the equivalent of a human running 200 miles an hour!

The smallest roach species grow to only 2.5 to 3.5 millimeters long – they belong to the genus *Ataphila* and they hang out with leafcutter ants, where they feed on the fungus that grows on the leaves the ants harvest. That's why they're commonly called ant roaches. There are nine species in this genus, and they're found from Louisiana in the U.S. all the way down to Argentina in South America.

The largest cockroaches can be over 4 inches long! There is some discrepancy when it comes to naming the largest cockroach species, though. Are we talking size? Weight? Wingspan? Size-wise, the *Megaloblatta longipennis* is the winner – it can grow to 4 inches, with a wingspan of around 7 inches. It's found from Mexico down to Peru.

Weight-wise, the *Macropanesthia rhinoceros* takes the cake – it clocks in at an average of 33 grams, but individuals have been known to weigh up to 50 grams. That's 1.5 ounces, or one-tenth of a pound… a lot for a bug! The common name for this cockroach is the Australian rhinoceros cockroach, the giant burrowing cockroach, or the Queensland giant cockroach. As the name suggests, it's found in the Queensland region of Australia, and it burrows into the ground. It's got front legs with several spikes that almost look like fingers extending from a hand. In fact, they've been likened to the paws of another burrowing animal – the mole!

The Australian roach can dig burrows up to 3 feet deep and over 20 feet long, where, unlike most other roach species, they will birth live young. Australia's giant burrowing cockroaches have another unique trait – they care for their young before they leave the burrow. That's uncommon for cockroaches and most insects in general. What's even more uncommon? Mama burrowing cockroaches look after their babies for up to 6 months before the little guys strike out on their own! In the insect world, that's devotion at the highest level.

Perhaps the most famous roach – one you've probably seen on TV or at the zoo – is the Madagascar hissing cockroach. Remember that cockroaches breathe through small holes in their bodies? These are found along the sides of their backs and their abdomens. That's how hissing cockroaches make noise. They don't do it with vocal cords but by breathing out heavily.

The air escapes so quickly that it makes a hissing sound (side note: Australia's burrowing cockroach also hisses!). Cockroaches don't have much in the way of defense mechanisms, except for their exoskeleton, so hissing is how these particular species scare predators away and communicate with their fellow roaches. It's also how males fight for territories and mates. Male hissers have horn-like structures protruding from their thoraxes, which they use to bash into each other as they hiss with fury.

My final fun fact about cockroaches is this: they can be trained! Have you ever heard of Pavlov's dogs? Pavlov was a physiologist who worked with dogs in the late 1800s and noticed that the dogs would salivate when they heard his assistant walking around just before mealtime.

So, he designed an experiment to test whether he could induce salivation by making other sounds before the dogs' food was delivered. Using a metronome, a clicking device popular with dog trainers today, he made clicking sounds before his assistant began preparing the dog food. At first, nothing happened since the dogs didn't associate the clicking sound with food. But after repeated attempts, the dogs *did* begin to associate the clicking sound with food. They'd begin to salivate when they heard the clicks, even when there was no food in sight.

This same type of conditioning has been done with cockroaches! Cockroaches pretty much universally dislike peppermint, but they *love* vanilla. So, researchers introduced both scents to a group of cockroaches. The vanilla scent was presented on its own, but the peppermint scent was presented with sugar water. It didn't take long for the roaches to begin associating peppermint with the sugar water. So, despite not liking peppermint, they'd congregate around that smell rather than the one they *did* like since it came with food. The cockroaches had been successfully conditioned!

These same researchers also discovered that a cockroach's ability to learn seems to be governed by its circadian rhythms. Those who underwent conditioning in the early morning retained hardly any memory of the association of peppermint with sugar water. Those trained in the evening, though, formed those associations quickly and retained their memory of it for much longer than the morning roaches. Essentially, roaches are bumbling idiots when they wake up and brilliant learners right before bed!

CHAPTER D
Dolphin
(**daal**·fn)

In 1968, Baba Dioum, a forestry engineer from Senegal, coined a phrase that I absolutely adore: "We will conserve only what we love; we will love only what we understand; and we will understand only what we are taught."

I think this statement is so true. People won't be motivated to care for the Earth and its creatures if they're never exposed to them. That's why, even though we try to limit screen time in our house, I always encourage Animal Planet, Discovery Channel, and National Geographic. When I was a little kid, and all my friends were watching Saturday morning cartoons, I was watching taped reruns of Shark Week. That's where my love of all things wildlife started.

Since becoming a wildlife biologist, I've had so many awesome animal encounters… I can hardly believe my good fortune! I'll tell you about my dolphin encounters in a minute, but first, let's dive in – pun intended! – starting with some facts.

Dolphins are awesome, and their amazing traits and characteristics are endless. First, there are 43 species of dolphins; the bottlenose is probably most familiar because it's the most common. There are three species of bottlenose dolphin: the common bottlenose, the Indo-Pacific bottlenose, and the Burrunan.

Bottlenose dolphins are one of the most recognized marine animals on the planet because of their use in the entertainment industry. Long before I ever saw a dolphin in real life, I was watching the one that starred in the TV series *Flipper*. (Yes, I'm *that* old!) The dolphin's actual name was Kathy, given to her by the trainers that prepped her for the show. Sadly, Kathy, as well as four other dolphins who starred in the show, were all taken from the wild to be used in the entertainment industry. Thankfully, that trend is mostly over; for the most part, wild marine mammals are no longer captured for the sole purpose of entertainment or captivity.

You're probably familiar with the fact that dolphins communicate, orient themselves, and find food by echolocation. If you've ever seen a dolphin in person or on television, those clicks that you hear them make are how they echolocate. These broad-spectrum sound waves reflect off nearby objects, sending information back to the dolphin about what the object is, whether it be a navigational hazard or a tasty fish.

Bottlenose dolphins, like their other ocean dolphin counterparts, catch food with dozens of sharp teeth and then swallow it whole. They hunt in groups, cooperating to locate and consume fish. The way they work together is an evolutionary wonder, perfected over thousands of years. When a pod locates a school of fish, they work together to herd the fish toward shallow water. The shallower the water, the harder it is for fish to escape; the harder it is for fish to escape, the easier they are to catch and eat!

One recently discovered feeding strategy has been dubbed "mud ring feeding." It's when a pod of dolphins locates a school of fish, herds it into shallow water, then "traps" it in a ring of mud. One dolphin swims in a circle around the school of fish while beating its tail on the bottom of the sea floor, which churns up a bunch of sandy, muddy sediment. This sediment rises around the school of fish, which the fish perceive as a barrier, and they essentially become trapped in this ring of sediment. Sometimes the fish will try to jump over the ring, only to have a dolphin catch them midair. Other times, the dolphins dive into the plume and chow down on the trapped fish!

Then, there's a phenomenon known as strand feeding. I've had the absolute privilege of witnessing a dolphin strand-feed on a saltmarsh bank in South Carolina. We're not talking shallow water hunting; we're talking *out-of-water* hunting. That's right; the dolphin was out of the water! Here's what happened…

I was prepping my kayak so I could go paddling for a couple of hours. I was planning to slide the kayak in the water on a muddy salt marsh bank when, suddenly, I heard some loud splashing. I looked up to see a dolphin swimming straight toward shore, creating a huge wake in the shallow water. Its body emerged more and more from the surface of the water as it approached the shoreline, then boom! A dozen fish leaped out of the water onto the mud, and the dolphin followed right behind. It beached itself, then thrashed around in the mud, grabbing the fish that were flopping around, before sliding back into the water, tail-first. The bottlenose did this several times as I watched in awe. It's right up there as one of my most thrilling life experiences!

The common bottlenose has one of the most well-developed intellects in the animal kingdom, right next to humans! They are also extremely social animals, again, much like humans. They communicate, play, and support each other in pods of over a dozen, with the majority of members being female or juvenile.

Speaking of playing, they don't just play with other dolphins – they play with humans, too! Another encounter I had, also in South Carolina, was when I was a counselor at a kids' kayaking camp. Now, when I took adults out paddling, we usually saw dolphins; there were so many in that particular area of the coast that it was unusual *not* to see dolphins when we were out on the water. It was always a thrill, especially if some came close, but we never really knew what we were going to get until we were out in the water.

With the kids, though, we were virtually guaranteed to see dolphins – and almost always up close! Now, this isn't because we were chasing after them, or feeding them, or any other things that are against the law. The dolphins would seek *us* out. There is no doubt in my mind it's because they knew I was with a group of kids. There's just something about a child's excitement that's contagious, right?

Well, no sooner than the kids would begin laughing and squealing as I taught them kayaking skills, the dolphins would show up in droves. They'd swim all around us in the shallow water, poke their heads up out of the water to watch us, and even vocalize like Flipper herself! It was an absolute thrill being on the water with those kiddos and witnessing how the dolphins responded.

And get this... dolphins give each other names. Seriously! Dolphins have been observed using very unique vocalizations when calling out to individual members of their pod. It was deduced that they were calling each other by separate and specific "names" as they communicated. Ever heard the phrase, "It takes a village..."? If dolphins could talk, they'd say it, too. They care for each other when one of them is sick, they assist when one is giving birth, and they even babysit for each other!

Males are a bit more solitary but join up with passing pods often, especially when it's time to mate.

On the subject of mating, females don't have it that great, and here's why. Males compete heavily for access to fertile females. This is usually in the form of fighting until the strongest male is victorious. It's also common behavior for a male to actually take a female dolphin hostage! He'll restrict her movement until she gives in and mates with them. This can sometimes take weeks! That said, sometimes dolphins will even exhibit mating behavior when females are outside their reproductive window.

Another cool thing dolphins do? They sleep with one eye open. Since they breathe air, they can't just zonk out completely, or they'd drown. So, they rest one half of their brain at a time, sometimes closing the corresponding eye, while the other half makes sure they're still propelling themselves through the water and surfacing every so often to breathe. They rotate resting each side of their brain for a total of about 8 hours a day. An 8-hour-a-day rest cycle... just one more trait that makes them so much like humans!

THE BONKERS BIOLOGIST

As I mentioned before, most people know about saltwater dolphins, but did you know there are also freshwater dolphins? And that they're *pink*?? At least, Amazon river dolphins are.

There are 5-ish river dolphin species alive today. Yes, that sounds weird, but it's hard to be exact because scientists are considering whether two sub-species have evolved to the point that they're two separate species. Maybe the exact number will be official sometime soon! Sadly, we already know that the baiji, or Chinese river dolphin, is likely extinct. They're the first dolphin species we know of that was driven to extinction by humans. I could go on a serious rant about humans' role in species extinction, but I'll save that for another book. After all, this one is supposed to be… amazingly amusing!

That brings me back to South America's Amazon river dolphins, also called botos. They're found in freshwater in 7 countries: Bolivia, Brazil, Colombia, Ecuador, Guyana, Peru, and Venezuela. Botos have more of a hump than a well-defined dorsal fin and can move their heads independently from the rest of their bodies. Now, that doesn't *sound* like a big deal… until you know that it's the only river dolphin that can do that! All other river dolphin species have fused vertebrae, so they are physically incapable of turning their heads without turning their whole bodies.

Unlike their marine counterparts, which move in large groups called pods, river dolphins are much more solitary. They're usually spotted alone, in pairs, or in very small groups of 4 max. Unlike other dolphin species that have sharp, conical teeth and swallow their food whole, botos have molar-shaped teeth and chew their food before swallowing it – like humans!

Botos play a large part in Amazonian culture, as they're often encountered by fishermen. They're playful and curious and have been reported to steal paddles from watermen canoeing along the river!

But the coolest part about botos? Their pink color. While they're born a typical dolphin-grey, their skin turns pastel pink as they age!

CHAPTER E
Electric Eel and Echidna

Electric Eel
(uh·**lek**·truhk **eel**)

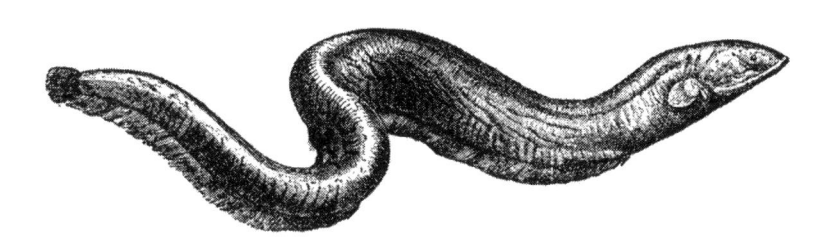

I'll start E with another animal I have a fun story about – the electric eel. But first, the facts…

Electric eels look much like marine eels, but they're more closely related to catfish than other eels. They live in the freshwater of the Amazon River and its tributaries, as well as smaller streams and ponds. They don't have pronounced scales like other fish; their skin has a smoother appearance, almost amphibian-like. If you ask me, they look kind of like salamanders – all the way down to their beady little eyes!

Don't let me confuse you, though – they aren't amphibians; they're fish. That said, they're an odd type of fish in that they don't breathe through their gills. They actually go up to the surface to breathe air! Many of the streams and ponds they live in are poorly oxygenated, especially in the dry season, so they've adapted to breathe in air through their mouths.

These bad boys can get pretty big, too – up to 8 feet long!

Now, the good part. These eels have three organs that work together to allow them to deliver electric shocks to anything – or anyone! – that bothers them. These organs are so large that they take up 80% of the eels' bodies. The other 20% houses all their other organs, all smushed together toward the front of their bodies.

These guys don't usually want to deliver large electric shocks because shocks drain so much of their energy. So, they'll try to just deliver enough for any given situation. For example, in addition to defense and predation, electric eels also use electric shocks to communicate with other eels. That would require mild electrical discharges. They also use electricity to find their way in the murky river water they inhabit. Another situation that only requires mild discharges.

On the other hand, when defending themselves or trying to take down large prey, the shocks will get bigger and more hazardous.

As I said, they don't want to just go around delivering electric shocks all the time. But, if they find themselves in a particularly perilous situation, they have been known to go out of their way to deliver a shock. For example, if an eel is trapped in a small pond or puddle with a large land animal hovering overhead, it can actually jump out of the water, fling its body onto the animal, and shock it before dropping back into the water!

Electric eels can deliver up to 800 volts if need be. That's enough to take down a human!

And that brings me to my story…

When I was in college, I interned at an aquarium, taking care of their bird, reptile, and insect collections. I was also assigned a special project – helping to construct the new Amazon exhibit. After weeks of building tanks and other enclosures, we got one of our first animal deliveries… an electric eel!

In order to transfer it to its new tank, two of our managers suited up in rubber coveralls and hopped in. As soon as they were both in, one commented that his feet felt wet. We asked if he needed to get back out and change into another set of coveralls, but he wasn't concerned. So, another supervisor handed the eel – which was in a smaller, portable tank – over to the managers, who gently lowered it into the new tank where it could swim out into its new habitat.

The transfer was successful! The first manager hopped back out while the other one – the wet one – waited for his turn. Well, gentle or not, successful or not, moving an eel is still a stressful event for the animal, and this eel wasn't settled yet. Before the second manager was able to leave the exhibit, he screamed and jumped into the air. The eel had discharged a shock… and in a twist of fate, the manager who said his feet felt wet had a small hole in his suit that had been letting water leak in. So, he got the full force of the eel's shock! In hindsight, I guess he should have gotten out first, or at least taken our advice to change into watertight coveralls. Oops!

Echidna
(uh·**kid**·nuh)

The echidna is one odd mammal. Mammals are characterized by being warm-blooded, having hair, producing milk, and giving birth to live young. The echidna mostly fits those, but with one glaring exception: it lays eggs!

Found only in Australia and New Guinea, echidnas are actually one of only two mammals that lay eggs. The other is the platypus, also found in Australia (spoiler alert: we'll be talking about the platypus in Chapter P!). Australia has plenty of interesting endemic creatures (side note: endemic means an animal only found in a particular region or country); these creatures have some unique adaptations for a unique continent!

Echidnas have quills like porcupines, pouches like kangaroos, and long snouts and tongues like anteaters! It's actually nicknamed the spiny anteater. But don't let that confuse you. They are not closely related to anteaters; they just capture prey in a similar way. Their sharp claws help them break apart logs and termite mounds so they can locate and slurp up ants, termites, and other insects with their long, thin tongues. They don't have teeth, but what they *do* have is really sticky saliva, and some have tooth-like spikes on their tongues. These sticky barbs act like Velcro, and insects stick right to them. Echidnas can eat a whopping 40,000 individual bugs each day!

There are four species of echidna in existence today. One species, Sir David's long-beaked echidna, is named after Sir David Attenborough! Don't know who that is? Well, have you ever watched a wildlife or nature documentary that was narrated by a man with a gentle British accent? Odds are, that was him. He's a famous naturalist and historian who's considered a national treasure in the United Kingdom.

Now, back to the echidna. The four species are categorized into two groups: short-beaked and long-beaked echidnas. They don't have actual beaks, so that classification is a bit misleading. Basically, we're just talking about short snouts versus long snouts. Depending on the species, echidnas grow to be between 1 and 2 ½ feet long and weigh between 5 and 36 pounds. Their quills can grow up to 2 inches long.

Remember that mammals have fur, right? Echidna's quills are actually large, modified hairs that have hardened. Underneath the quills, they have shorter fur all over their bodies that keeps them warm. When it gets too hot out, though, echidnas must burrow or hunker down in the shade. They lack the ability to sweat at all, so they can overheat if the temperature gets too high. Echidnas have low body temperatures of around 89 degrees, one of the lowest of any mammal on earth. They also have super slow metabolisms. This allows them to be pretty long-lived for such small mammals – echidnas in captivity can live to be 50 years old!

Early in their evolution, it is thought that they may have been venomous because they still have genes that contain traces of poisonous peptides (side note: peptides are amino acids made by the body). They also have spurs on their legs, like the platypus. Male platypuses use their spurs to deliver venom, but echidnas are no longer venomous. Scientists think the male's venom evolved over time into non-poisonous secretions that are released during mating season when they're trying to attract the ladies!

When males *do* come across a potential mate, they follow her around, as do other interested males, until she allows the dominant male to mate with her. For some weird reason, the several males who end up competing for her will form a single-file line as they follow her movements. Researchers have dubbed this behavior the "echidna train!"

Once the female echidna is ready to mate, the train then becomes a sumo wrestling match! She finds a comfy place to lay down, then the males dig a trench into the ground, completely encircling her. The strongest male then shoves the other males out of the circle until he's the only one left. That makes him the winner and gives him the honor of becoming her mate. After mating, the female lays a single fertilized egg in her pouch, where a tiny echidna hatches after about ten days. It stays in the pouch for another couple of months while it grows and its quills harden, then emerges to live in the burrow its mom dug.

Baby echidnas drink their mom's milk until they're about seven months old, then head off to establish their own territories. They don't nurse like other mammals, though. Mama echidnas secrete milk through pores in their bellies; so, instead of breastfeeding, babies lap up the milk like dogs drinking from a puddle!

Echidnas have fairly poor eyesight, but they make up for that by having the unique ability of electrolocation (note that *electrolocation* is different than *echolocation*). While common in aquatic animals, only four terrestrial animals have this ability! Echidnas sweep their noses in a left-to-right motion over the ground – like you might see someone doing with a metal detector at the beach – and thousands of sensory receptors in their snouts are able to detect electrical signals produced by the prey underneath the soil. They then use their powerful claws to dig up the prey before slurping it down!

My final fact is the most insane of all: echidna evolution dates back so far – to the era of dinosaurs, actually – that it's thought to be earth's first mammal!

CHAPTER F
Frog
(fraag)

When it comes to listing the amusing facts about frogs, it's hard to know where to start. With over 4,800 species of frogs on earth, a whole litany of unique abilities and behaviors have been observed. For example, did you know that frogs can't throw up? Weird! What if something they eat makes them sick?

When humans eat something bad, our stomach's way of protecting us is to reject the food by contracting until we vomit. But what about frogs? Let's say they eat a poisonous insect. How do they get rid of it so they don't get sick? Well, instead of just vomiting out their prey, they engage in gastric eversion.

Sounds kinda scientific, right? Let me give it to you in layman's terms... gastric eversion is just a fancy phrase for puking your guts out – literally! That's right, frogs eject their entire stomach out of their mouths, turning it inside out in the process. The prey falls out, the frog uses its front feet to scrape off anything that's left behind, and then swallows its stomach again!

Wait, there's more... some frogs that live in climates characterized by extreme cold weather can actually freeze solid and then "come back to life" when temperatures rise again. Talk about crazy! When human cells and tissue freeze solid – which can happen to fingers and toes, or even entire limbs, in cases of extreme frostbite – we lose the limbs that froze. This is because the freezing process sucks all the water out of our cells, causing them to collapse. Once they reach that point, no amount of thawing or rehydration can plump the cells up again. They're dead, and the tissue must be removed.

That's not the case with one special category of frogs. There are at least six species that can survive being frozen. Known species are: in the Southern hemisphere, New Zealand's brown tree frog; in the Northern hemisphere, specifically North America, Cope's gray tree frog, the eastern gray tree frog, the spring peeper frog, the western chorus frog, and the wood frog.

The wood frogs of Alaska have been well-studied with regard to their freezing-thawing abilities. Here's what happens: once the first ice-crystal appears on a frog's skin, it triggers the liver to release a ton of glucose (essentially liquid sugar) into the cells. As the frog's blood begins to freeze, the sugar doesn't, so the cells stay plump and hydrated. This protects the frog's major organs, which sustain little to no damage from the freezing event. When the frog thaws, its body picks up where it left off, and the heart and organs kick back into gear. This can happen several times over the course of the cold season, and the frogs are no worse the wear from it!

On the flip side, some desert-dwelling frogs can reanimate after prolonged drought. "Drought" and "frogs" aren't typically words you hear together because frogs, like most amphibians, need moisture to survive. Let's focus on the arid regions of Australia's central and western deserts. Burrowing frogs are found in these deserts, and, as the name suggests, burrowing is a key aspect of their survival. A couple of species are especially adept and can remain burrowed in the dry, hard ground for years. Yes, *years*!

Here's how they do it: when a good rain saturates the area, they use their large hind legs like shovels, digging backwards into the ground to make a hole about 3 feet deep. They settle in, excreting a cocoon-like layer of dead skin cells around their bodies to keep moisture in, and fall into a deep sleep. While in this hibernated state, they don't require the energy reserves needed when they're awake and active. So, their metabolism slows to the point that it nearly stops, allowing them to remain dormant for years without losing much energy. After all, they need it for when the next rain comes so they can dig back out of the newly-softened ground and look for a mate!

After emerging, the clock is ticking down the minutes until the next dry spell, so burrowing frogs have to take care of this mating thing pretty quickly. Males find the closest temporary pond and perform their mating calls to lure in the ladies. Once mating is complete, females lay up to 500 eggs in the ponds, where they need to mature before the water dries up. Mother nature helps make that possible by gifting burrowing frogs with a relatively quick development – it only takes about a month for tadpoles to become frogs. Then, they need to eat like crazy to fatten up in time to bury themselves.

Their parents will have already been working on that, eating insects and even cannibalizing their fellow frogs as they race against the clock before the next drought. If the ground dries up too much before burrowing frogs have built-up their fat stores, they won't be able to dig through the hardened mud to hibernate. They'll dry out and die.

Burrowing frogs are hard to study since they emerge so infrequently. But scientists estimate some can live up to two decades. In that time, it's not uncommon for them to only come to the surface 3 or 4 times. Can you imagine? Only seeing the daylight four times in 20 years?!

CHAPTER G
Galápagos Tortoise and Greenland Shark

Galápagos Tortoise
(guh·**laa**·puh·gows **tor**·tuhs)

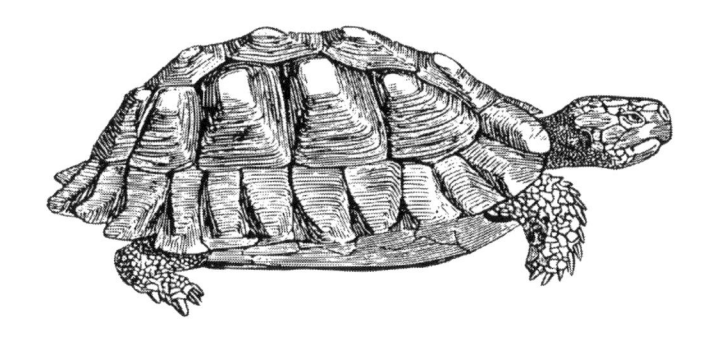

G alápagos tortoises are one of the longest-lived terrestrial animals on earth (terrestrial means they live on land). They can live up to 150 years and weigh in as much as 500 pounds! Weird for an animal that can go a whole year without food or water. That's right, an entire year! Their bodies have a unique ability to store food and water reserves to get them through dry seasons.

Let's compare… on average, humans can only live for a few days without water. The Guinness world record holder went 18 days without water. Going without food for a while isn't so bad (the world record for a human is 380 days – please don't attempt to set a new record for this!), but a whole year without water? Very, *very* few animals can do that and survive.

The Spanish word for tortoise is galápago, which makes sense since Spanish explorers are thought to have been the first humans to discover the islands in the 1500s. At that time, they encountered tens of thousands of tortoises, hence the Galápagos Islands' name.

Because they're so long-lived, Galápagos tortoises don't reach mating age until around 20-25 years old. Like most reptiles, Galápagos tortoise females aren't very maternal. They lay their eggs in burrows about a foot deep and then leave. When the babies hatch, they have to dig out of the burrow before completely fending for themselves on land as brand-new hatchlings.

While only about 12 inches underground, it can take a month for the hatchlings to dig their way to the top to emerge. They start slow and spend the rest of their lives moving slow, too. Let's compare again… a Galápagos tortoise's average walking speed is 0.16 miles per hour. A human's is 2.8 miles per hour. That's over 17 times faster! These guys aren't in a hurry.

Galápagos tortoises live a life of leisure. They sleep for about 16 hours a day and spend their waking hours sunning themselves and eating low-lying vegetation. Like I mentioned earlier, Galápagos tortoises are pretty slow-moving animals. That said, during mating season, males perk up quite a bit. When two are vying for the same female, the competition is fierce. And by fierce, I mean fierce by Galápagos tortoise standards! Males will stare each other down before rising up on all fours as high as they can and stretching their necks as high as possible (pretty intense, right?). The one who stands tallest wins, and the defeated male usually draws his head back into his shell to acknowledge defeat. As I said, it's a pretty fierce battle (is my sarcasm coming through well?)!

Depending on the island they live on, the tortoises' bodies have adapted to suit that particular island's food source. On islands covered in grassy pastures, tortoises have domed shells. This restricts upward head movement somewhat, but that's not a problem when their food sources are low to the ground like they are. On islands with sparser ground-level vegetation, the tortoises have saddleback-shaped shells that allow for upward movement of the head, so they can reach higher up to eat taller-growing plants, bushes, and cacti.

Have you ever seen these guys walk? Probably not in person (if you *have*, I envy you!), but maybe on television? If not, head on over to YouTube and look it up. It's so darn cute! Their front feet are turned inward, so they don't exactly get from point A to point B in a straight line. Instead, they heave their shells left and right with each step in what looks like a slight swagger. So cute!

Greenland Shark
(**green**·luhnd **shark**)

The Gs take the cake for the oldest animals on the planet – at least, the oldest vertebrates (we'll talk more about that in Chapter H!).

The most famous Greenland shark, a 16-foot female whose eye lenses were carbon dated by Danish scientists, was found to be *at least* 272 years old. At most, she'd be 512 years old. *Five hundred twelve*! What?! Yes, I know that's a pretty big gap, but carbon dating can give a range as opposed to an exact date. But even on the low end at 272 years, that's an insanely long lifespan!

Like Galápagos tortoises, Greenland sharks are also extremely slow-moving animals. Kinda makes you think, doesn't it? Maybe if we humans just slowed down once-in-a-while instead of being go-go-go all the time, we'd live longer, too. Or at least be more content?

It's their slow-moving nature, combined with a slow metabolism, that allows Greenland sharks to live so long. They grow so slowly – only a centimeter a year – that they don't even reach mating age until about 150 years old. Can you believe that? No human that we know of has ever lived past 122, but Greenland sharks don't even start having babies until they're 150!

Even more amazing than that, if you can even imagine, is that mama Greenland sharks carry their fertilized eggs for a minimum of 8 years before laying them. What the heck, right? Well, read more closely… I said a *minimum* of eight years. The maximum gestation period is thought to be 18 years. *Eighteen years*! Greenland sharks can be pregnant for the same amount of time it takes for a human to be born, grow up, and move out of the house!

That's about all we know about Greenland shark reproduction. Mating has never been observed in the wild, so no one knows where it happens, what time of year it happens, if there are any cool mating rituals… these living fossils are just too elusive.

Moving on from baby-making, another thing we *do* know, though, is that several of the Greenland sharks that scientists have studied have been blind. And it wasn't a natural occurrence. Yes, it's true that these sharks live at such depths that they're in complete darkness, so their sense of sight isn't very useful anyway. Their eyes are actually very small – millions of years of evolution in an environment where sight wasn't a primary means of finding food or mates mean that their eyes didn't need to be very large or complex. So, they gradually got smaller over time.

But evolution didn't cause them to completely lose their sight – at least, not yet. What researchers found in every single Greenland shark they studied was the presence of parasites attached to their eyeballs. They're called *Ommatokoitas* and look like small, pinkish-white worms. Ew!

Ommatokoitas attach, sometimes permanently, to the corneas of the shark and eventually destroy the tissue, rendering the shark blind. Weirdly, it doesn't seem to cause the shark pain, and even the blindness it causes doesn't seem to affect the lifespan of the shark. Since Greenland sharks don't rely on their eyesight anyway, they just go about their lives with worms hanging off their eyeballs for hundreds of years!

CHAPTER H
Hydra and Horseshoe Crab

Hydra
(**hai**·druh)

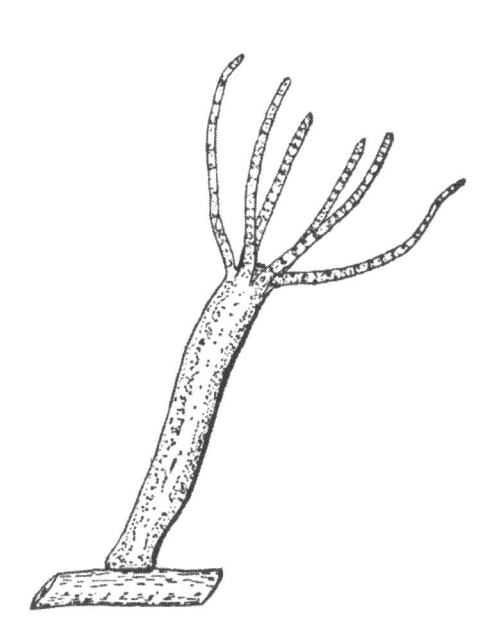

While we're on the subject of long-living animals, how about I throw in one that's immortal? The hydra is believed to have no natural lifespan... as in, unless it's killed by something, it never dies!

Put simply, hydras are freshwater polyps. Polyps are very small organisms related to jellyfish, anemones, and coral. Hydras have a tubular form with skinny tentacles at one end, all packaged in a body less than half an inch long. The tentacle-less end is the one that attaches to substrate (pssst… substrate is just a scientific term for the surface that an organism attaches to).

Like many aquatic species, hydras can regenerate pieces of tissue if they're injured, just like the serpent in Greek mythology. The mythological creature was said to regrow its head each time it was cut off. Hydra polyps can do the same thing, except with their entire bodies!

Unlike most other animals, a hydra's body is primarily made up of stem cells, which continuously divide and differentiate throughout its life. Differentiation is key – it means the new cells can replace old cells in any part of the hydra's body where fresh cells are needed. So long as this process continues, hydras never age or die.

But does this process truly continue in perpetuity? In a 4-year study that took place in the 1990s, researchers wanted to know whether, and at what rate, hydras aged. They did this by studying the timeframe within which individual hydra lost fertility and eventually died.

The hydras did neither. Researchers found absolutely no evidence of aging in hydras, leading them to conclude that, in ideal conditions, hydra can live forever!

Obviously, hydras in the wild don't usually live in ideal conditions. They face predation, pollution, and a number of other conditions that just about every other wild animal in the world faces.

So, the vast number of hydras don't live forever. But just the fact that an animal has that ability is wild!

Let's compare… human embryos lose their full-body stem cells after just a few days of development. In adulthood, stem cells are only found in humans' bone marrow. But if scientists can figure out the specifics of how hydra stem cells operate, they may in the future be able to transfer that knowledge into therapies that help with healing human injuries, especially degenerative diseases.

Thanks, hydra!

THE BONKERS BIOLOGIST

Horseshoe Crab
(**hors**·shoo **krab**)

Have you ever heard the term "living fossil?" That's how horseshoe crabs are typically referred to. And, boy, do they look the part! With a spiked, 2-part, dome-shaped exoskeleton covering their bodies, and a long, pointed tail protruding from their domes, they definitely look like something from another era. In fact, they've been in existence in their current form for over 350 million years!

But there's much more to horseshoe crabs than just their age and unusual appearance. First, did you know they're not true crabs? They have ten legs like crabs do (more on that in a minute), but evolved from an ancient group of invertebrates more closely related to spiders and scorpions. Like spiders and scorpions, they have several pairs of eyes. The weird part, though? Horseshoe crabs not only have eyes on their heads but also one beneath their tails. I'll say it again... horseshoe crabs have an *eye* on the underside of their *tails*!

These eyes can do some pretty cool things, too. Not the "tail eye" so much; it's not as intricately developed as the "head eyes." The most noticeable pair of eyes is on the front section of the dome, called the prosoma. That's the part that looks like a horseshoe – hence, the crab's name. These two compound eyes are highly developed, with around 1,000 receptors each. They are very sensitive to changes in light, adapting quickly at dawn and dusk to see very well in both bright light and darkness.

There are five more sets of eyes on the prosoma behind the 2 compound eyes. These bad boys detect UV light from the sun, as well as moonlight. Scientists have determined that these eyes function to help horseshoe crabs detect new and full moons, so they know when it's time to spawn.

The single eye on the tail (proper name telson) is composed of several photoreceptors that further help their brains determine day and night.

The horseshoe crabs' teeth are also in a weird place – their feet! What is the *deal* with this animal?! Okay, okay, let me clarify. Their mouths don't have any jaws or teeth, yet they eat food that needs to be chewed. To do so, they use spines at the base of their legs to mash up their food before sucking it up into their mouths. So, they basically chew their food by walking all over it!

I mentioned that they have ten legs, but if we're being technical, they actually have 12 appendages – 6 on each side. The first five sets are the legs that have those chewing spines for crushing food, as well as claw-like "feet" at the ends that they use to move along the ocean floor. The last pair, appendages 11 and 12, don't have claws but rather paddle-like structures. These are used to shovel sand so they can burrow into the sea floor, much like other crabs do.

On top of all these unique physical features, there's another hidden feature that's of particular use to humans – horseshoe crab blood! Horseshoe crabs have contributed more to our medicine and medical research than any other marine animal. Their blood contains a special clotting agent that's activated in the presence of toxins and germs. When bad bacteria are present in a horseshoe crab's body, its blood clots around the bacteria, effectively isolating them so they don't spread. One of the many ways pharmaceutical companies use horseshoe crab blood is to test vaccines and drugs before they're distributed to make sure they don't contain contaminants. If the blood clots, impurities are present; if it doesn't, the medication is good to go!

One last cool detail about horseshoe crab blood? It's high in copper, so it turns bright blue when exposed to oxygen. When researchers draw horseshoe crab blood, the labs are filled with gallons and gallons of electric blue liquid gold (literally, horseshoe crab blood can fetch prices that rival the price of gold)!

Since horseshoe crabs have a hard exoskeleton, they must molt as they grow. They molt, meaning they shed their old shells, almost 20 times before they reach maturity. Once they've finished their final molt, they're ready to spawn.

Speaking of spawning, did you know that horseshoe crabs are the primary supporter of North America's red knot bird population? *Wait, what?* What do red knots have to do with horseshoe crab spawning?

Red knots are small birds that spend winters in warm climates from North Carolina to the tip of South America, then migrate to the Canadian Arctic to breed in the spring. Horseshoe crab eggs are critical to their survival during migration because they need an abundant food source to sustain them on their long journey up the coast of the Atlantic Ocean – some fly as far as 9,000 miles on their migration from Brazil to Canada! So, they time their migration with horseshoe crab spawning season, making sure they stop right when there are plenty of eggs to gobble up.

The main stopover point for the red knot is Delaware Bay, the location of the largest population of horseshoe crabs on Earth. Thousands upon thousands of horseshoe crabs emerge from the water each year in May and June to lay eggs in the sand. This is where those special eyes, which help horseshoe crabs determine lunar cycles, come into play – they can sense the new and full moons, which result in especially high tides, and that's when they emerge to lay their eggs. It is a unique and exhilarating sight to see all those horseshoe crabs gathered in one spot along the shoreline!

A single female can lay 100,000 eggs over several spawning days – add that up over several thousand crabs, and that's hundreds of millions of eggs each season! More than half of the red knot population that migrates each year stops at Delaware Bay to fatten up on horseshoe crab eggs before resuming their journey to Canada. They eat so many of these eggs – several thousand, in most cases – that they can double their body weight in a matter of days before departing!

Surprisingly, that doesn't hurt the horseshoe crab population at all. What can I say? Nature is smart! There's always a balance, and, in this case, the crab population continues to thrive while also providing life-dependent sustenance to these red knots – they simply wouldn't survive their migration journey without the spawning horseshoe crabs.

CHAPTER I
Indian Giant Squirrel and Indian Palm Squirrel

Indian Giant Squirrel
(**in**·dee·uhn **jai**·uhnt **skwur**·uhl)

There are a dozen squirrel species found in India, each with its own unique characteristics. Here, we'll focus on the Indian Giant Squirrel and the Indian Palm Squirrel, obviously because they start with the letter "I."

Like all 200 plus species of squirrels, Indian squirrels are in the Sciuridae family, along with prairie dogs and chipmunks. Squirrel species are further categorized into three types: tree squirrels, ground squirrels, and flying squirrels.

The Indian giant squirrel is a tree squirrel that spends most of its life high in the treetops of India's rainforests. Their tails are longer than their bodies, and they're used expertly for balancing up in the trees, especially after a long jump. The Indian giant squirrel can leap up to 20 feet at a time!

And it's far larger than most squirrels – in fact, it's one of the largest squirrels that exist today. Let's compare…

The grey squirrel, the most common squirrel in the eastern U.S. and one that I see several times a day, averages 21-30 inches long (tail included) and weighs 1-1.5 pounds. The Indian giant squirrel, on the other hand, averages 26-42 inches long, weighing in at 3.5-4.5 pounds!

That's on par with what is currently recognized as the largest squirrel species in the world – the Laotian Giant Flying Squirrel, which is averages 42 inches long and over 4 pounds. That said, this is a relatively newly-discovered species, having been first observed in 2013. And it wasn't seen by a bunch of researchers in the forest in Laos; it was discovered in a cage at a meat market! After that, biologists set out to find it in the wild, but it's only been seen a handful of times. So, it's hard to be sure what their average size truly is.

Giant black squirrels also rival the Indian giant squirrel in terms of length – they can reach up to 43 inches long. But they usually top out at a much lower weight than the Indian giant squirrel, at max 3.5 pounds.

Unlike many squirrels that store nuts and seeds underground in preparation for winter, the Indian giant squirrel stores its food high up in the tree canopy. They like to build several nests in their territories – I like to think of it like humans having different rooms in their houses. Who doesn't want a little variety once in a while? Sometimes I want to lounge on the couch, sometimes on the back porch, sometimes in my bedroom. Maybe Indian giant squirrels want a different view depending on their mood, too. Who knows?

And it is gorgeous! Unlike the plain greyish-brown squirrels we are familiar with here in the U.S., the Indian giant squirrel has multiple colors. It's nicknamed the "rainbow squirrel," and its coloring has even been compared to technicolor! It has various hues of brown, white, and black, combined with the brighter colors of red, maroon, and even pink!

The various colors help them to camouflage in the trees to avoid predators, where they lie flat and still against tree branches. Yes, I know pink isn't typically a color of camouflage in a forest, but we're not talking hot pink here. It's more of a dark fuchsia. It, along with the other light and dark hues, blends into the vegetation, shadows, and dappled sunlight that filters through the forest canopy.

THE BONKERS BIOLOGIST

Indian Palm Squirrel
(**in**·dee·uhn **paam skwur**·uhl)

Are you familiar with chipmunks – like maybe *Alvin and the Chipmunks*? In the U.S., these rodents have distinctive stripes on their backs. An Indian palm squirrel looks a bit similar, with three white stripes running down its back. Even though babies are born hairless, the white stripes are still visible on their skin.

Though sacred in India, the palm squirrel is considered a pest in other parts of the world where it doesn't belong. For example, Australia. There are no squirrels native to Australia, but, back in the late 1800s, a small population of palm squirrels was introduced to the grounds of the South Perth Zoo in Western Australia. Park employees thought they'd be a fun addition for visitors to see and interact with while visiting.

Not surprisingly, the squirrels – like most unenclosed animals – didn't acknowledge the zoo's perimeter as their holding area, so they expanded past the boundaries of the zoo. An exclusion zone was created to prevent further expansion outside Perth, but they roamed beyond that, too. Unfortunately, an extermination program eventually had to be put in place. The latest figures show the palm squirrel at around a dozen individuals.

Non-native species typically damage the natural ecosystem in areas where they've been introduced, and the palm squirrels in Australia are no different. They opportunistically prey on bird eggs, which put native nesting birds under threat. Even at such a small size, they pose a significant risk to local wildlife, as well as crops.

At 11-15 inches long (including the tail), they are on the small side for a squirrel. And they don't even weigh a pound; they barely weigh a quarter of a pound! On average, they weigh in at 0.2-0.25 pounds. Just imagine the difference if you had an Indian palm squirrel and an Indian giant squirrel side-by-side!

Many tree squirrels hibernate, but not the palm squirrel. Instead, it snuggles up in its nest and waits out the cold, relying on its food stash to make it through to the warmer months. Since it doesn't hibernate, it relies on a stable food supply during winter. As such, it will aggressively defend its cache and will vocalize a loud chip-chip-chip sound if a predator or competitor is nearby.

THE BONKERS BIOLOGIST

Like I mentioned previously, the Indian palm squirrel is considered sacred in its native India. In old Hindu stories, a palm squirrel helped a deity, called Lord Rama, to build a bridge across the ocean to rescue his wife. Afterward, when the deity stroked the animal, his fingertips left stripes down its back!

CHAPTER J
Jerboa
(jr·**bow**·uh)

Jerboas are so stinking cute! Imagine a small rodent with a mouse-like body, face, and tail, the ears of a jack rabbit, the hind legs of a kangaroo, and gigantic eyes relative to their size. That's what jerboas look like!

There are 33 species of jerboas, all belonging to the family Dipodidae, which is exclusive to jumping rodents. Jerboas sometimes walk on all fours but are primarily seen balancing or hopping on their hind legs because their forearms are relatively short – again, like a kangaroo. Jerboas can jump up to 5 times the length of their bodies! They range in size from the super tiny Baluchistan pygmy jerboa to the great jerboa.

The Baluchistan pygmy jerboa, also called the dwarf three-toed jerboa, looks like a little cotton ball with a tail. No kidding! When at rest, it looks like a round ball of fur – something a cat might throw up!

Unlike other jerboa species, a pygmy jerboa's ears aren't crazy long. They're a fairly normal size relative to the rest of its body. Speaking of, a pygmy jerboa's body grows to just over an inch and a half, with a tail length of about 3 inches. It weighs in at a tenth of an ounce – that's a hundredth of a pound! According to the *Guinness Book of World* Records, this jerboa species is tied with the African pygmy mouse for the world's smallest rodent.

Remember the hissing cockroaches we talked about earlier in the letter "C?" They weigh ten times more than Baluchistan pygmy jerboas. Imagine, an inset that weighs ten times more than a rodent!

The great jerboa is found in the 'stans (aka. Kazakhstan, Turkmenistan, and Uzbekistan), as well as in parts of Russia and Ukraine. They grow up to 18 inches long (tail included). Great jerboas dig both permanent and temporary burrows. They have two permanent burrows, one for summer and one for winter, plus temporary ones that they occasionally use to rest.

Long-eared jerboas are 9-10 inches long (including their tails), so smack in between the Baluchistan pygmy jerboa and the great jerboa, size-wise. Its ears are about 30-35% longer than its head, hence the name. Long-eared jerboas are found in the Gobi Desert of Mongolia and China. Because of their limited range, they're culturally significant in the area. The long-eared jerboa is even featured on a limited-mint Mongolian coin – a 500 tögrög silver coin with a Swarovski crystal insert for the jerboa's eye!

If you ever come across any jerboa species, best not to mess with them. Although they're adorable and non-aggressive toward humans, scientists are pretty sure they're carriers of monkeypox, a virus that's a relative of smallpox. Do a Google search… you DO NOT want this disease! In fact, the U.S. has banned the import of all African rodents because of the threat of monkeypox. Ick!

Jerboas are desert-dwellers that share a fascinating trait: they don't drink! How on earth does a desert animal survive without drinking water? Well, the jerboa gets all the hydration it needs from its food… plants and insects, to be exact. They're so specially adapted to desert conditions that not only can they do without water, but they can also withstand temperatures as low as freezing and as high as 130 degrees Fahrenheit.

Jerboas rely on their jumping abilities to escape predators. One species, the rough-legged jerboa, has been observed leaping up into low-hanging vegetation, grabbing onto leaves, and running up into the branches to avoid predation.

Once jerboas are born, they won't reach maturity for over three months. That doesn't sound long, but it's actually about twice as long as most other small rodents!

Okay, final fact about jerboas. Their leg bones bear a striking resemblance to bird leg bones. Like birds, several jerboa species have tridactyl feet, meaning only three toes. Scientists think this may indicate convergent evolution, whereby completely different species that occupy the same environment evolve similar physical and behavioral traits to survive. Very Cool!

CHAPTER K
Kookaburra
(**koo**·kuh·br·uh)

Kookaburras are interesting-looking birds. At least, they are to people like me who don't live in Australia! They're a species in the kingfisher family and have relatively stout beaks for a bird. They've got pretty neutral brown and off-white coloring but for a few bright blue feathers on their wings.

Unlike other kingfishers, kookaburras don't hunt fish. They eat land animals, small and large, including snakes, lizards, other birds, and small rodents. If small enough, the kookaburra will just swallow its prey whole. If it's on the large side, though, you may want to cover your eyes!

I had the privilege of studying abroad in Australia in my 20s and was fortunate enough to have several kookaburra encounters. There are two I'll never forget. The first was when I saw a kookaburra perched on a branch with what looked like a mouse in its beak. The kookaburra beat the living daylights out of the mouse before consuming it, repeatedly slamming it against the branch until it was limp as a ragdoll. It was just too big for the bird to swallow without smashing it to bits first. It was pretty fascinating… and gross!

My second encounter was on a camping trip in Grampians National Park. But first, a couple more kookaburra facts…

While the majority of flying birds migrate for the winter, Kookaburras are one of several species that don't. Instead, to survive the cold, they engage in communal roosting, snuggling up together to keep warm.

Researchers have observed that non-migratory birds have more aggressive behavior and are bolder and more curious than their migratory counterparts, especially when it comes to foraging. When the seasons change and prey becomes more scarce, non-migratory birds need to expand their palates. I witnessed this firsthand on my camping trip!

Let me set the stage for you… one weekend during my semester studying in Australia, several students and I decided to go camping in Grampians Park. We went on some great hikes, saw awesome waterfalls, kangaroos, the whole nine… but that's another story.

One day, we were having lunch at a picnic table at our campsite. As we were prepping and setting out food, a very curious kookaburra was flying all around above us, trying to find the right opportunity to steal something. There were several of us getting ready for lunch, though, and he wasn't brave enough to dip right into the middle of the group. Or so we thought…

He got braver and braver (we knew it was a "he" because only males have blue feathers on their tails), and the longer he flew, the lower he swooped each time he took a pass at the picnic table. After the kookaburra chickened out several times, imagine the hilarity that ensued when the person sitting across from me picked up his hotdog only to have that bird zoom in from the side, grab the sausage mid-dive – right out of the bun in my friend's hand! – and then fly off into the tree canopy! That kookaburra didn't miss a beat, and we at the table absolutely died laughing! So, I guess bravery could be listed as one of the kookaburra's amazing traits!

That wasn't the end, though… you see, we'd just finished a long hike before returning to the campsite for lunch, and we were all sweating, including one of my friends who wore hearing aids. Sweat was dripping over the top of his ears, so he decided to remove his hearing aids so they wouldn't get too wet. He took them out and put them on the table.

Despite what we'd just witnessed, none of us had the foresight to consider what might happen to my friend's hearing aids when he tossed them in the middle of the picnic table. Yep, you know where this is going… no sooner had they landed on the table than the kookaburra swooped back through, grabbing one and carrying it off! We all sat there with our mouths hanging open as my friend – almost in slow motion – leaped up from the table yelling, "NOOOOOOOOOO," as the bird flew off into the distance.

And that was the day I learned just how expensive hearing aids are!

CHAPTER L
Loggerhead Shrike and Lobster

Loggerhead Shrike
(**laa**·gr·hed **shrike**)

During one of my first jobs after getting my master's degree in wildlife management, I worked on a captive breeding and release program for a sub-species of loggerhead shrike.

Shrikes are passerines of the taxonomic order Passeriformes, as are about half of the bird species on the planet. Passerines, simply put, are small perching songbirds.

One of the Channel Islands off the coast of California is called San Clemente Island (or SCI), which is home to a sub-species called the San Clemente Loggerhead Shrike. This sub-species is endemic to SCI, meaning that it only exists on this island.

The first humans to arrive on SCI were ranchers in the mid-1800s. They brought cattle, sheep, and goats with them to graze the island. There were lots of bushes, scrub, and cacti on SCI, but not a ton of trees, so the grazers cleared much of the island's vegetation pretty quickly. Sadly, the shrike population started to decrease due to habitat loss.

The other thing humans brought to the island? Cats and rats. So, the taller trees and vegetation that didn't get cleared by the grazers were all that remained for the birds to nest in… and for the cats and rats to climb. So, in addition to habitat loss, the shrikes also had to deal with nest predation from newly-introduced feral animals, in addition to their natural predators – namely, the island fox.

This story sounds depressing, but there's a happy ending! Things turned around when the U.S. Navy took ownership of the island in 1934. The ranchers left with most of the cattle and sheep but didn't take the goats, and they continued reproducing until there were nearly 12,000 of them! So, in the 1970s, the Navy got to work removing them. It was a long effort, but by the mid-90s, the last goat was finally gone!

Then, the Navy got to work again, this time funding several conservation organizations to replant vegetation and enhance the shrike population by implementing a captive breeding and release program. I had the insane privilege to work on the release effort, where I got the opportunity to introduce captive-reared birds into the wild. I observed shrikes both in captivity and in their natural habitat and saw some pretty amazing and unique behaviors.

Let's start with the mating behavior of the San Clemente loggerhead shrike. I'll admit, it isn't unique to shrikes, but I have to talk about it because it's just so darn cute and because I witnessed it firsthand!

When a pair would take a liking to each other, the male would show his interest by offering the female food. The female would show *her* interest by performing this type of "damsel in distress" behavior. She'd perch on her branch and flutter her wings as if she was trying to fly but couldn't. She'd accompany this with a pathetic call as if she was crying for help. She'd flutter and call as the male searched for food and, when he found a tasty morsel, he'd bring it to her and put it in her mouth. Manipulation at its best!

Or maybe this is just how the ladies would prepare their men for fatherhood… after their hatchlings arrived, the little guys displayed the same begging behavior whenever mom or dad would return to the nest with food. Either way, it was fascinating to see this mating and child-rearing behavior up close!

Now, on to feeding. You know how I mentioned that kookaburras kill their prey pretty violently? Well, shrikes have pretty grisly feeding behavior, too. They eat several prey items, from insects to small reptiles to small rodents. When they catch their prey, if it's small enough, they swallow it down whole. If not, they shake it ferociously until it dies or is knocked unconscious. Then, because they

THE BONKERS BIOLOGIST

don't have talons sharp enough to hold the animal while they tear it into smaller pieces to eat, they impale it on a cactus thorn or sharp twig. There it hangs while the shrike rips pieces off to eat. I've seen everything from crickets to mice impaled near shrike nesting areas. Creepy!

Now, back to the captive breeding program. I'm happy to report that it's been pretty successful! In 1998, the population consisted of only 14 individual birds. By 2009, the population was almost 200 individuals! I was so excited and honored to have played a small role in the program… and to have been paid to basically do my hobby every day (hiking and birdwatching) wasn't too shabby either!

Lobster
(**laab**·str)

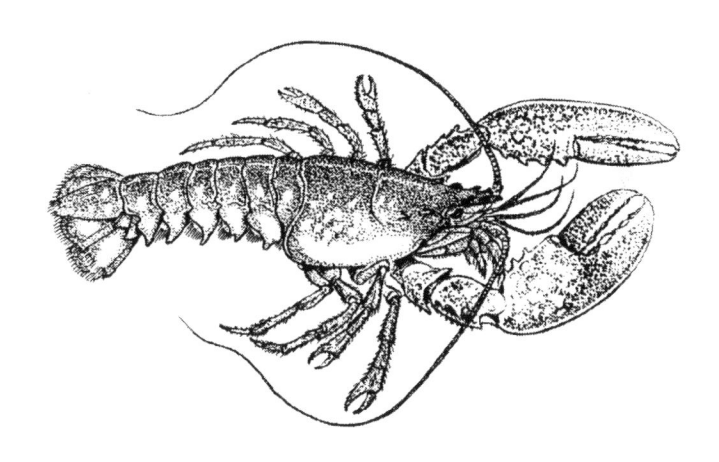

There are so many cool animals that start with L that it was difficult to narrow down my list of which to feature. The leafcutter ant and leaf-tailed gecko are two of my favorites, but in the end, I'll tell you about an animal you've already heard of – the lobster. But what you maybe *haven't* heard of is the lobsters' unique anatomy.

Their brain is in their throat instead of their head, they taste with their feet instead of their tongue (they don't even *have* tongues), and their teeth are in their stomach instead of their mouth! There's more… let's dig in.

There are about 75 species of lobsters in the world, and they grow continuously throughout their lives, unlike many animals that stop growing after reaching maturity. In fact, a lobster holds the record for the largest crustacean ever known – a little guy (errrr... big guy) that was caught in 1977 off the coast of Nova Scotia. He weighed in at 44 pounds 6 ounces and was thought to have been 100 years old!

Researchers believe that lobsters have relatively poor vision, though they aren't entirely certain to what extent. Many think that lobsters can only detect light and shadows, as opposed to complete images, and they don't see the full color spectrum. So, they rely on sound, taste, and smell to move through the environment. But they don't hear with ears or taste with mouths or smell with noses – all these are accomplished via their legs and antennae.

A lobster's large antennae have tiny sensory hairs called mechanoreceptors that detect sound. These receptors don't actually "hear" as humans do, but rather they identify vibrations from the sounds that prey, predators, and other lobsters make in the water.

A lobster's small antennae have hundreds of tiny hairs that do the smelling. These hairs can detect the levels of various compounds in the surrounding marine environment – especially amino acids given off by other animals. By sniffing around (aka. flicking their antennae), lobsters can tell if there are predators or prey nearby and in which direction they're located.

Finally, a lobster's legs have more sensory hairs that taste potential food items as the lobster walks over them. If it's suitable for eating, and if it's soft, the lobster will use its smaller claw, called the pincher claw, to tear pieces off and put them into its mouth. If the prey is hard, like a mollusk inside a shell, the lobster will use its large claw, called the crusher claw, to break it apart before consuming what's inside. A lobster's claw can close with the force of 100 pounds per square inch!

Once swallowed, prey goes directly to the lobster's stomach, where it's chewed. You heard that right; lobsters chew their food *after* it enters their stomachs! Inside their stomachs is what's called a gastric mill, where several plates called ossicles grind up their prey until it's small enough to pass down into the gut.

Question: What happens to our food and drink after it's processed by our digestive system? What goes in must come out! So, let's talk about pee… most animals release urine from their backsides. Not the lobster, though – they pee out of their eyes! Technically, it's from pores just under their eyes, at the base of their antennae. Also, unlike most animals, lobster pee isn't just a waste product. It's how they communicate!

While lobsters can make various rasping and popping sounds with their bodies, they can't vocalize because they don't have vocal cords. So, they communicate by squirting pee in each other's faces. That's too funny not to repeat… they pee in each other's faces! Their urine contains chemicals that signal various messages to other lobsters – that they recognize them ("Hello, neighbor!"), that they're protecting their territory ("Beat it, trespasser!"), or that they're ready to mate ("Hey there, cutie!").

And finally, a lobster's brain is not located in its head, like most species. Instead, it's in their throats. Don't ask me why… maybe it's because their kidneys and bladders take up too much room in their heads?

Lobsters are typically a brown-ish color – it isn't until they're cooked that they turn red – but rare and fascinating color variations

THE BONKERS BIOLOGIST

have been observed by lobstermen in their catches. Thankfully, these lobsters usually escape their fate on someone's dinner plate since lobstermen usually donate them for display. As of 2021, the best place to see these color variations in person is at the New England Aquarium in Boston. They've got a blue lobster, an orange lobster, a yellow lobster, and even a half-and-half – literally, a lobster whose left half is black and whose right half is orange!

All of these are super rare – 1 in 2 million for blue, 1 in 20 million for orange, 1 in 30 million for yellow and calico, and 1 in 50 million for an even color-split right down the middle. The rarest color variation, an albino lobster, is a 1 in 100 million anomaly!

CHAPTER M
Mantis Shrimp
(man·tuhs shrimp)

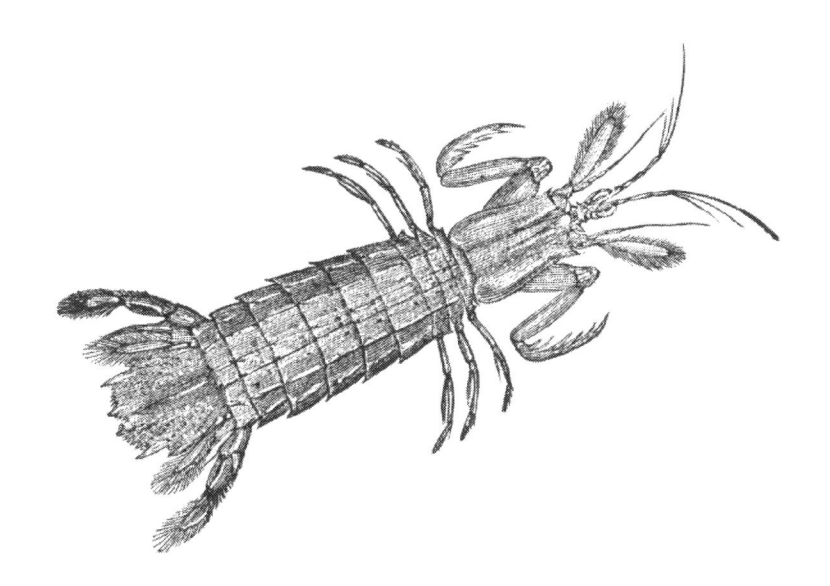

Mantis shrimp have the fastest punch in the animal kingdom. How fast? So fast that it produces enough heat to boil the water around it! The mantis shrimp's punch creates so much energy so quickly – in 3 thousandths of a second, to be exact – that the water around it temporarily reaches the temperature of the sun!

Let's back up… first, mantis shrimp aren't true shrimp. Yes, they look like shrimp or small lobsters, but they're actually pretty distant relatives of those animals. They've evolved so much in the past 400-ish million years that they now belong to an order all their own: Stomapoda. There are over 550 species of mantis shrimp in this taxonomical order.

Did you notice I mentioned that mantis shrimp have been around for over 400 million years? That's before dinosaurs came along!

Okay, back to the mantis punch. If you've ever seen a praying mantis insect, you may have noticed that they hold their two front legs in a folded position in front of their bodies like they're praying – hence, the name. Mantis shrimp have a similar posture, with two arm-like appendages folded just under their heads. Mantis shrimp use these appendages to punch their tasty and unsuspecting prey with such force that it pretty much pulverizes them. If it doesn't kill them, it definitely knocks them out! Sooooo... never cross the path of a mantis shrimp?

These punching appendages take one of two forms – spears or clubs. Some mantis shrimp have barbs on the ends of their arms that spear prey and predators when they punch. Others have clubs on the ends of their arms that smash prey and predators when they punch.

Club-fisted mantis shrimp are the fastest punchers, so let's talk about them. Their punch is so fast that it can't be seen by the naked eye... it can't even be caught on film with standard high-speed video equipment. Scientists have had to use super-specialized high-speed video cameras to capture a mantis shrimp punch, then slow the footage down over 800 times to be able to view it. What they discovered was a punch that reaches 50 miles an hour in three-thousandths of a second, so fast that a flash of light is emitted when the punch lands, causing the water around it to momentarily boil!

Mantis shrimp don't punch the way we do. When humans punch, we cock our arm back and use our arm muscles to force our fist forward into our punching bag. Their arm works more like a spring-loaded apparatus. When it's cocked, a spring-like structure in the joint is compressed while a cog-like structure locks the arm in place. The upper arm muscle builds up more and more energy as time goes by. When the mantis shrimp needs to punch, whether for predation or protection, the cog releases, the spring expands, and all the built-up energy shoots the lower arm out like a bullet. Literally. A mantis shrimp's 50-mile-an-hour punch accelerates faster than a bullet exiting a gun!

This force is more than enough to knock a fish dead, crack the shell of a crab, and even break glass. It's hard to keep mantis shrimp captive for this very reason. They've been known to shatter aquarium glass. They've even split the flesh of people! Yep, unfortunate mantis shrimp owners have had their fingers cut to the bone by a single punch! Same with fishermen!

But what are the drawbacks? Does it hurt a mantis shrimp when it cracks bones or shatters glass? Nope! Its exoskeleton is like armor. It's made up of some seriously strong crystallized calcium, with layers of chitin – a flexible, fibrous material that acts as a shock absorber – underneath. It's so sophisticated that researchers have modeled new carbon fiber body armor after it.

This extra protection comes in handy for mantis shrimp, which fight much more often than your average animal. Mantis shrimp are more aggressive than other crustaceans because they're all in competition for the best places to live and, once they find that perfect crevice, they're very territorial of it. So, they don't just save their punches for food and protection; they also beat the crap out of each other as they compete for shelter!

Before moving on, I'll cover one more trait that makes mantis shrimp amazingly unique – they growl! Males emit a low-pitched rumble when guarding that territory that was so hard to come by, letting others know they'll have a fight on their hands if they don't stay away. It appears that mantis shrimp are also much smarter than the average crustacean because they seem to remember individual shrimp they've done battle with. Scientists have observed that mantis shrimp will avoid adversaries they've lost punching matches to, as well as their territories.

It's thought that the mantis shrimp's growl also attracts mates – and once the ladies choose their guy, some mantis shrimp species stay together for life; not all species, but some. Monogamy is super rare in the crustacean world!

CHAPTER N
Nudibranch
(**noo**·dee·braank)

Nudibranchs are, hands down, the most beautiful animals in the world (I mean, just look at the book cover)! I'll admit, that's just my opinion… but surely you agree! Nudibranchs are ocean-dwelling mollusks with over 3,000 species in existence today. They have colorations ranging from plain white with black spots, to dark green with lime green spots, to off-white with purple spots and stripes, to purple with yellow spots and stripes, to solid purple with neon orange accents, and the list goes on! Again, I recommend doing an internet search. Right now. The images will amaze you!

Nudibranchs are mollusks, like slugs and snails. Mollusks don't have bones, just soft gelatinous bodies. They're usually covered with a shell, with a couple of exceptions. Slugs are mollusks but don't have shells. Nudibranchs don't have shells either. That's why they're commonly called sea slugs.

Wait. Let me back up. So, nudibranchs *do* have shells briefly during their larval stage. But the shells are shed after a few weeks, during their metamorphosis into adult form. Mature nudibranchs are hermaphrodites, meaning they have both male and female reproductive parts. They still need a mate to produce fertilized eggs, but after finding a partner and mating, *both* individuals will lay eggs.

Nudibranchs produce long, thin egg sacks that look sort of like ribbons. They lay these egg sacks near corals or sponges or anemones, in the hopes that they blend in with their surroundings enough that predators won't notice them. For unknown reasons, most nudibranchs lay their egg sacks in a clockwise spiral. Intriguing!

With more than 3,000 species, there's a lot of variation in nudibranchs. They range in size from less than ¼ inch – about the length of a kid's fingernail – to over a foot long. They can weigh between a few ounces to over 3 pounds. They all have sensory organs on their heads that look like a pair of antennae. These vary in size, color, and texture but are used for the same purpose – detecting food, mates, and predators.

Nudibranchs are carnivores, eating other animals in the tropical reefs they inhabit. They'll chow down on fish eggs, coral, sponges, anemones, and… even other nudibranchs. Yep, these guys eat their own – they're cannibals! However, some nudibranchs get their energy from photosynthesis. Wait, isn't that a plant thing? Why, yes, it is! Some species eat corals that are rich in algae, then store the algae in their bodies where it photosynthesizes, providing the nudibranch the energy it would normally derive from its prey.

And here's a fun fact: nudibranchs get their wild, vibrant coloring from their food. Just like flamingo feathers are pink because of the shrimp they eat, nudibranchs absorb the pigment of their prey, too. That coloring is then transferred to the nudibranchs' own skin.

One of the ways scientists have kept nudibranchs organized is by dividing them into two groups based on their gill structures. Dorid nudibranchs breathe through a plume of gills on their backs right by their rear ends – it sort of looks like a tiny anemone is hitching a ride on their butts!

Aeolid nudibranchs breathe through cerata, which are horn-looking structures found all over their bodies, not just a single tuft of gills like dorids. The cerata provide respiration but also defense. This is because they can store poison from other animals. Most nudibranchs don't produce poison themselves, but if an aeolid nudibranch eats a stinging animal that contains toxins, it can absorb that poison into its body, then direct it to the tips of its cerata where it can poison any predator that tries to take a bite!

I mentioned earlier that nudibranchs live on tropical reefs. That's mostly true, but there are always exceptions to the rule. Nudibranchs are no different! There are a few species that live in the frigid waters of the Antarctic, some that live on the ocean floor thousands of feet deep, and some that spend their lives swimming through the water column or even at the surface. This handful of exceptions is specially adapted to those different environments.

Take the blue glaucus nudibranch, for example. Rather than living on reef structures or the sea floor, they spend their lives swimming throughout the water column, feeding on the tentacles of Portuguese man o' war jellyfish. Despite the vivid blue color on their undersides and gray coloring on their backs, they avoid detection by swimming *upside down*. Any predators in the water below just see gray, which is the color the ocean surface looks like from underneath. Any flying predators in the air see blue, which is the color of the water from above. So, they usually don't notice the blue glaucus and fly right by. What a cool adaptation!

CHAPTER O
Opossum and Oyster

Opossum
(uh·**paa**·sm)

Y ou may only know opossums from seeing them as roadkill on the highway. But these elusive animals have some awesome survival adaptations (at least, when cars aren't involved!).

But first, the facts... opossums are marsupials – mammals that birth underdeveloped babies, then carry and nurse the babies in a pouch on their bellies. They are the only marsupials found in North America. Compare that to the almost 250 marsupial species in

THE BONKERS BIOLOGIST

Australia! And here's a fun fact: North American opossums have 50 teeth – that's more teeth than any other North American mammal!

Historians think the word opossum originated from the native American word "apousoum." It's a Powhatan word that means "white dog," and the word was first recorded in the writings of John Smith, the English settler who helped found the Virginia colony of Jamestown. You probably know his name from stories of Pocahontas.

Though the word was first written in the early 1600s, opossum fossils have been dated back to dinosaur times – the oldest on record is from 70 million years ago! Scientists think marsupials evolved from placental mammals (aka. mammals that give birth to fully-developed young) starting around 125 million years ago.

Okay, on to the fun stuff... opossums are really smart! Researchers have conducted learning and cognition tests on opossums and have found that they outscore man's best friend – yep, opossums are smarter than dogs!

Opossums have opposable thumbs on their feet – all four of them! That, along with a prehensile tail, helps them masterfully climb trees and hang on to branches for added stability. It also helps baby opossums, called joeys, hold on to mom when she's carrying them around. Once they emerge from her pouch, at about four months, they climb onto her back, riding along as she explores and forages.

When a predator poses a threat, the first thing an opossum will do is rise up on its hind legs and hiss loudly and continuously, in the hopes the predator will get nervous and back down. The thing is, though, opossums are all bark and no bite. If the predator continues to advance, the opossum won't charge and fight. Instead, the opossum will pass out, making the predator think it's dead.

The thing with predators is that, for the most part, they like to hunt and kill their food. If it's already dead, they're not as interested – that's where scavengers come in. Since they're not picky, scavengers will happily eat dead things, even if they've been rotting for days. Ick! Not predators, though. If they come across a dead animal, they'll leave it alone the majority of the time and go in search of live prey.

But I digress. The thing about this "passing out" behavior in opossums is that it's completely involuntary! That's right; opossums don't decide if or when to play dead. Rather, their body's natural defense mechanism takes over and knocks them unconscious. They awake a few minutes to a few hours later and go about their business as if nothing happened!

But every once in a while, they can still get a nasty bite from a predator. If it's a snake, though, that's where their next amazing ability comes in handy... opossums are immune to most snake venom! A bite that could seriously poison or kill another animal is typically no match for the opossum. In fact, researchers believe that it would take at least ten venomous snake bites to make an opossum even a little woozy.

Opossums also appear to have *some* immunity to rabies – they have far fewer infections than other animals in rabies-infested areas.

An opossum's primary prey is insects, but they also consume rodents like rats and mice. They aren't picky! If you have opossums on your property, don't try to get rid of them. They are your natural pest control!

Opossums also eat reptiles, birds, eggs, plants, trash... pretty much everything. This includes carrion – aka. dead animals. When they aren't roadkill themselves, you'll often see opossums *eating* roadkill, which means they'll probably *end up* as roadkill eventually since they're literally feasting in the roadway. I guess they'll never escape the stereotype!

Oyster
(**oy**·str)

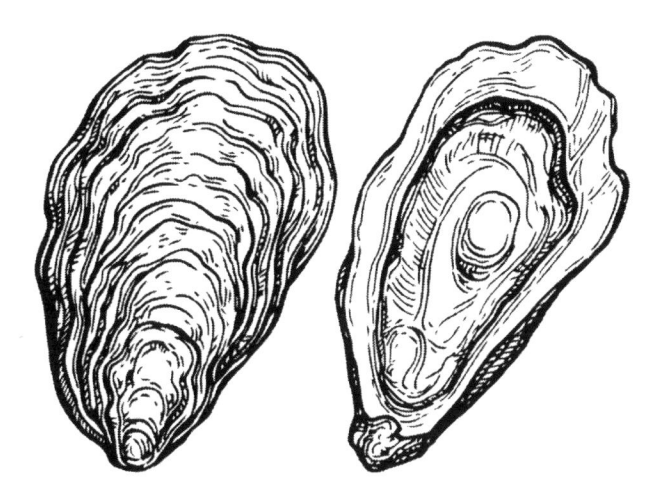

I love oysters! And not just for eating (actually, I don't have a taste for them, so don't eat them at all), but for their massive role as ecosystem engineers. These guys can congregate by the tens of thousands to create reefs that purify water, provide habitat, and protect shorelines.

Oysters are filter feeders. They suck in water, filter out the food, then release the water back into the sea. But oysters don't discriminate. In addition to food, they also filter out dirt and pollution. So, the water they release is clean, making oysters nature's natural water purifiers. A single oyster can filter 50 gallons of water per day!

THE BONKERS BIOLOGIST

After graduating from college, my first job was in Hilton Head, South Carolina. I was employed as a naturalist and hosted kayak nature tours for visitors. Remember my dolphin encounters I mentioned earlier? Well, I also saw a ton of oysters too. The marsh was covered with little pockets of oyster reefs that would become visible as the tide went out. It was such fun seeing them emerge because, as the water receded, you could actually see the oysters releasing the water they were filtering.

Usually, you can't observe oysters filter-feeding because – duh! – they're underwater. But, just as the intertidal area starts to be exposed when the tide is going out, you can see it. And talk about amusing. When an exposed oyster releases filtered water, it "spits" the water straight up into the air! Imagine hundreds of oysters shooting water out of their shells as their reef emerges – it's like watching a choreographed performance at a water park splash pad!

Oysters also provide habitat for small animals. Their reefs can span a few feet wide to several acres. An oyster reef has so many nooks and crannies that are perfect hiding places. So perfect, that many aquatic species use oyster reefs as nursery areas. They deposit their eggs in the reef where they're protected from predators, and when the babies hatch, they can feed off of nutrients in the sediment while staying hidden from larger fish.

And finally, oysters provide shoreline protection from erosion. After my stint as a naturalist, I was privileged to work with oysters a second time while employed as a field biologist. Here's the backstory… back in 2010, the oil spill at the Deepwater Horizon drill site off the coast of Louisiana resulted in 4 million barrels of oil being released into the Gulf of Mexico. British Petroleum, aka. BP, was deemed responsible for the damage, and, in 2012, they reached a criminal settlement that required them to pay out $4 billion for habitat restoration.

At that time, I was working as a fisheries conservation manager at a large nonprofit organization, and part of my job was to allocate those settlement funds to different restoration projects in the Gulf. Much of the funding went to the building and rebuilding of oyster reefs. Eventually, I actually moved to the Gulf myself and was able to work directly on several restoration projects building oyster reefs in Mobile Bay off the coast of Alabama.

The projects were primarily intended to protect vulnerable shorelines from erosion. When reefs are present near the shore, they help to reduce wave energy that would normally wash out the sandy sediment. With oyster reefs, sediment is actually allowed to build up between the reefs and the shore, not only stopping further erosion but actually reversing it. It was a thrill, and so very fulfilling, to be able to contribute in a hands-on way to restoring a small corner of the Gulf of Mexico.

And now, on to the sad part… since the early 1900s, populations of oysters have declined more than 85% globally. They were once so abundant that people didn't think twice about harvesting them. But they've now been harvested to the point of near-collapse. Let's look at the Chesapeake Bay on the east coast of the United States, the largest bay in the country. It contains around 19 trillion gallons of water. There were once so many oysters there that they could collectively filter all the water in the bay in a single week! Today, filtering that much water takes over a year. That's a lot of oysters gone.

Thankfully, there are several organizations dedicated to the mission of restoring oysters. Oyster harvests are now heavily managed, reefs are getting rebuilt, and oyster spat (aka. baby oysters) are being captive-bred. I had the privilege of visiting a breeding facility when I was working for the non-profit.

The University of Maryland runs an oyster hatchery that has released over one billion oyster spat into the Chesapeake Bay since 2010. Breeding occurs in tanks in the lab, and newly-hatched oyster spat are allowed to grow for a couple of weeks before being transferred to huge open-air tanks on a pier at the water's edge. These tanks are filled with oyster shells – the oyster spats' preferred substrate – and the spat will attach to an available shell. Once settled, the spats' shells harden over several days, and then, they're ready to be relocated to their new home somewhere in the bay!

There are other restoration efforts underway that you can actually be a part of, too. Several seafood restaurants have gotten involved in oyster restoration by keeping the shells of the oysters they serve. Then, once a week or so, the left-over shells are collected by organizations like the ones I used to work for, tied up in mesh bags, and deployed to reef sites. Since oyster shells are the best substrate for attracting oyster spat, these shells will be the foundation for future oyster reefs. You can help this effort by patronizing seafood restaurants that donate their oyster shells for reef-building projects!

CHAPTER P
Platypus and Paddlefish

Platypus
(**pla**·tuh·puhs)

In my opinion, the existence of the platypus is a sure sign that God has a sense of humor. Think about it… a mammal with a duck bill? Ridiculous! Here's the thing about mammals; according to the rules of taxonomy, mammals are animals that have warm blood, bony skeletons, hair or fur, give birth to live young, and make milk for nursing those young.

There aren't really any exceptions to the "birthing live young" part – except for two. One, the echidna, we've already talked about.

THE BONKERS BIOLOGIST

The other is the platypus. These are the only two mammals that lay eggs!

Let's back up for a second. Taxonomy is the science of classifying animals and plants based on shared traits and physical characteristics. There are seven taxonomic levels. From the broadest to the most specific, these levels are: kingdom, phylum, class, order, family, genus, species.

The platypus has the physical characteristics of several different animals. It is so unique that it's not only the sole member of its genus and species but also the sole member of its family. Being the only animal in 3 taxonomic ranks is super rare.

Let's look at the platypus from head to toe to see why. The platypus is semi-aquatic and can spend about 2 minutes underwater swimming and foraging for food. Its duck-like bill has nostrils that can be sealed while it's submerged. It has webbed feet similar to an otter which helps make it an efficient swimmer, and a tail like a beaver which acts as a rudder, helping it twist and turn effortlessly while swimming.

Male platypuses have an extra feature on their hind feet: venomous spurs that deliver a toxic sting to any predator that dares mess with them! This is also pretty rare in mammals; of the approximately 6,400 species recognized today, only a couple dozen have the ability to deliver venom. The platypus used to be hunted for its fur and, while its venom isn't potent enough to kill a human, plenty of hunters were temporarily incapacitated by it before the platypus fur trade ended. These days, platypus stings are uncommon in humans.

The female's unusual feature is that it lays eggs – it bears repeating that platypus and echidnas are the only mammals on earth that lay eggs instead of birthing live young! When the eggs hatch, though, after about a 10-day incubation period, normal mammalian nursing occurs. Mama platypuses nurse their young in underground burrows for around four months before the nestlings emerge almost fully grown.

Babies are born with teeth inside their duck-like bills but, when they're ready to start eating solids, they lose their teeth. Wait, isn't that backward?! It certainly is for most mammals. We humans are born toothless, and our teeth grow in when it's time to start solid food. But the platypus *loses* its teeth when it's time to start solid food? I'm not saying Mother Nature doesn't know what she's doing, but I think we can all agree that this defies logic… at least, *our* logic!

That said, the platypus doesn't need teeth as an adult. No, platypuses don't swallow their food whole; instead, they scoop up small rocks and gravel into their mouths that they use to pulverize prey before swallowing it. Platypuses have hard pads in the back of their mouths made of keratin. They grind the gravel-laced food against these pads to aid in the pulverization process.

Yet another amusing platypus fact? They glow! Yep, when their skin is exposed to UV light, their fur emits fluorescent shades of purple, blue and green. This isn't that uncommon in the animal kingdom, but it is for mammals. How cool!

THE BONKERS BIOLOGIST

Paddlefish
(**pa**·duhl·fuhsh)

I had to feature the paddlefish because of my own encounter with one while I was working in South Dakota as a wildlife biologist. I've got a lot of stories, don't I? What can I say? Being a field biologist has its perks!

Several years ago, I was surveying and banding piping plovers nesting along the Missouri River. We used a little skiff to cover as much ground – I mean, river – as possible and were always encountering fishermen. The river was especially full of carp. We knew this because, whenever we pulled up to the shoreline or a sandbar and accidentally herded them into shallow water, they'd leap out of the river to get away. They'd woosh past our heads as they jumped over the boat, and we'd have to duck like we were playing dodgeball. Sometimes they missed the water and actually landed *in* the boat!

We'd always have a good laugh, after we made sure no one got nailed, of course. But, while I knew the river was filled with carp and other fish, I had no idea it also had paddlefish. Actually, I didn't even know what a paddlefish *was* until I saw one washed up on a sandbar.

Imagine my astonishment when I walked up on a whopping 6-foot-long fish with a giant paddle-shaped snout! Talk about someone's jaw hitting the floor. And that's not even as big as they get. The largest paddlefish ever landed in the U.S. was 7.1 feet long and weighed just shy of 200 pounds!

Up until the early 2000s, my Missouri River paddlefish – proper name "American paddlefish" – had a cousin: the Chinese paddlefish. It was even larger than the American one and could reach a max length of 23 feet! The record length caught by a fisherman was 9.8 feet, and it weighed over 660 pounds!

Sadly, the Chinese fish was declared extinct recently, with the last confirmed sighting having been in 2003. It was lost because of overfishing and habitat fragmentation from dam construction. We humans must do better! That leaves the American paddlefish as the only living species of a fish that's been around for over 120 million years. In the very river basin where I was working, the Missouri River, 60-million-year-old paddlefish fossils have been found!

Enough about fossils and extinction. On to more fun facts. This big fish also has a big mouth that can open disturbingly wide! Not to worry, though… if you're ever splashing around in a river that has giant paddlefish in it, you won't be bitten. Paddlefish are filter feeders. That giant mouth sucks in water, then the gills separate all the plankton and small organisms before forcing the leftover water out again. The paddlefish gulps down the rest!

Like the extinct Chinese paddlefish, the American paddlefish also faces threats. Besides direct human threats, a big one is the zebra mussel. Zebra mussels are an invasive species in U.S. rivers. They originated in Russia and spread throughout Europe and then the Americas by hitching rides on cargo ships. They are very efficient feeders and breeders, and can wipe out food sources that other animals rely on.

But wait, how can a tiny mussel threaten a giant fish? Through competition for resources. Like other bivalves, zebra mussels are filter feeders. And they're so good at it that they can completely clean the water of plankton before other filter feeders are able to consume enough. In that way, they pose a threat to paddlefish survival. Crazy how something so small could eat so much to pose a threat of starvation to something so big!

CHAPTER Q
Queensland Grouper
(kweenz·land **groo**·pr)

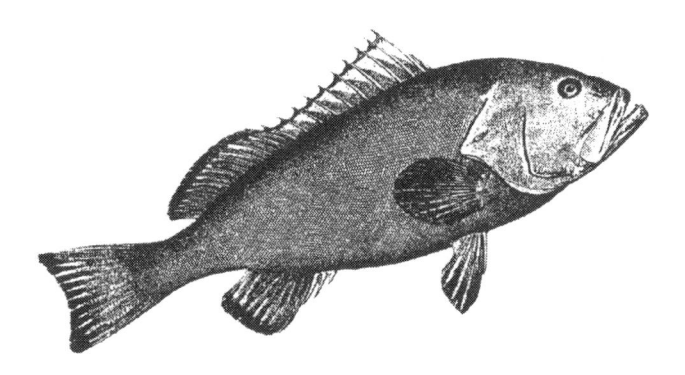

T he Queensland grouper is yet another animal I have a story about! But first, the facts… it grows to almost 9 feet and tops out at almost 900 pounds. In fact, it's the world's largest bony fish that inhabits coral reefs, almost exclusively found in shallow waters around reefs and shipwrecks in the Pacific Ocean.

Sounds ominous, right? Well, it's not. These guys are so slow-moving, they might as well be slugs. In fact, a lot of the time, they're spotted resting completely still – sometimes in the middle of the water column and sometimes on the ocean floor. They're also pretty cautious and would prefer to swim away from trouble rather than hold their ground. They're only the slightest bit aggressive and fast-moving when lunging at their prey. They don't bite their prey, though. They catch it by suction, taking in a huge gulp of water along with their meal.

On the subject of meals, Queensland groupers eat all sorts of typical prey: small fish, crab, and lobsters. They also eat baby sea turtles and... wait for it... sharks! Yes, they're so big they can take on small sharks. Even though they swallow their prey whole, they do have several rows of teeth. These are used to prevent prey from escaping.

Another weird fact about Queensland groupers is that they're always born female. As they age, some change to males, but the majority remain female to ensure mass reproduction. There only needs to be a few males to fertilize a lot of eggs, but there needs to be a lot of females to lay a lot of eggs. That's just how number games work. The more eggs there are, the more babies hatch. The more babies hatch, the more individuals will survive to adulthood. The more that survive to adulthood, the more eggs are laid. And the cycle continues!

Remember back in the kookaburra section, I mentioned that I spent a semester studying abroad in Australia? Well, after the school semester ended, I did some traveling before flying back home to the U.S., and one of the places I visited was the Great Barrier Reef. I mean, you can't go to Australia and *not* see its most famous natural wonder, right? It's literally listed as one of the seven wonders of the world!

Luckily, my sister was able to fly to "Oz" to do some of that traveling with me. We rented a motor home and drove up the east coast, stopping at cute little coastal towns as we made our way north to Airlie Beach. From there, we booked a snorkeling trip to the Great Barrier Reef. Our boat took us out, the instructor got us suited up, and into the ocean we went. Over 20 years later, I still vividly remember the vibrant corals, giant clams, and… the biggest fish I've ever seen in my entire life!

My sister and I looked at each other wide-eyed through our diving masks as we watched this enormous gray fish swim ever-so-slowly through the water around the reef. It must've been hundreds of pounds. It was a dull gray with bulging eyes and a huge underbite… it was a Queensland grouper! After we marveled at it for a bit, we turned our attention back to the reef and the colorful fish darting through the sponges and anemones.

We finally turned back to the open water to make our way back to the boat. And what did we come face-to-face with? That gigantic grouper, swimming straight toward us. Now, remember, these are super slow-moving fish. He was moving at a snail's pace. But when you have a fish *that* big heading straight for you, it's more than a little unnerving.

My sister and I grabbed hold of each other like we were facing certain death (spoiler: we were *not* facing certain death), and each screamed at the top of our lungs, spewing massive amounts of bubbles from our snorkels until we didn't have any air left. Then, we each tried to push the other in front of the fish so we could get away! I guess that sisterly bond goes only so far, huh?!

We popped out of the water and ripped our masks off, laughing and squealing until we were blue in the face! And that's the story of the day I was almost swallowed by a Queensland grouper (again, kidding… we weren't in any danger!).

CHAPTER R
Ruby-Throated Hummingbird
(**roo**·bee **throwt**·duhd **huh**·ming·burd)

There are over 330 species of hummingbirds in the world, all found in the western hemisphere. Actually, *only* found in the western hemisphere. No hummingbirds exist in the eastern hemisphere. And the ruby-throated hummingbird is the only one that nests in the eastern U.S., where I live. So, they hold a special significance for me!

Hummingbirds got their name because of the humming sound their wings make when they flap. At more than 50 beats per second, we humans can't see much more than a blur. But we can definitely hear it! Their wings beat so fast that hummingbirds are able to reach speeds of 40 miles an hour. They can also hover in midair, and they're the only bird species that has that ability. They can even fly backward and upside down!

A hummingbird's legs are short and weak compared to other birds. So weak, in fact, that hummingbirds can't walk. The best they can do is perch and occasionally shuffle sideways on branches.

Like all hummingbirds, ruby-throats are tiny. They measure 2.5-3.75 inches long – about 1/5 of their length is in their beak – and top out at 1/10 of an ounce... that's six thousandths of a pound! They're so tiny that their eggs are about the size of a jellybean. Just imagine how tiny their babies are when they hatch!

Ruby-throated nests are made of the usual plant material, with one additional special ingredient: spider silk. It allows the nest to expand as the babies hatch and grow. Cool, right? And get this... when females are pregnant, their legs can swell – just like humans! Ruby-throats typically lay two eggs, but nests have been observed with as little as a single egg and as many as three eggs.

To call them "ruby-throated" is a bit misleading because only the males have ruby-colored throats. Females typically just have white feathers in the throat area.

Most ruby-throats don't live past five years, but one bird that had been banded by researchers at birth was recaptured six years and 11 months later – it's the oldest on record! That's the beauty of banding... you can get exact ages to the day. While most hummingbird species are thought to live a max of 5 years, the oldest tagged hummingbird to ever be recaptured was almost double that – 9 years and one month old!

Here's how hummingbird banding works: first, the birds are either caught in traps containing nectar or in a mist net. Mist nets, made of thin mesh, are suspended between two trees like a volleyball net. Mist nets are strategically placed in whatever area has a researcher's target bird, and then they simply wait for birds to fly into the net. Traps are preferred for hummingbirds, but mist nets don't discriminate, so if a hummingbird flies into one, it'll get stuck. Afterward, the researcher removes the bird from the trap or net, takes weights and measurements, and records any other observations. Then, a small metal band – sort of like a tiny bracelet – is placed around the bird's leg and given a slight squeeze to close it up tight.

Bird bands have unique numbers printed on them so individual birds can be identified if they're recaught at a later date. Since hummingbirds are so tiny, so are their bands. They can easily close around a safety pin! After being banded, the hummingbird is released back into the wild.

Ruby-throats migrate from the Caribbean in the summer, and the vast majority of them are banded during this migration period. Their destination is the eastern U.S. and Canada; then they return to the Caribbean for winter. That's a roundtrip of up to 4,200 miles for the Canadian nesters! In order to make such a long trip – sometimes, with only one stop along the way – they eat a lot ahead of time. They put on an extra 30% of bodyweight before they start their trip to get them through the journey.

Their preferred diet is nectar from red and orange flowers. Hummingbirds are unique in that they can digest a lot of sugar. Most birds wouldn't survive on nectar because their bodies don't have the ability to process all that sugar. Hummingbirds are also unique because of their specialized beaks, which are long and thin – perfect for inserting into the openings of narrow flowers.

Contrary to what most people think, hummingbirds don't use their beaks like straws. Rather than sucking up nectar, they use their tongue to lap it up, sort of like a dog at a water bowl. Their tongues are forked at the end and have small hair-like protrusions that help the nectar stick to it. The difference between a hummingbird and a dog is speed – hummingbirds can lap up nectar as fast as 20 times a second. Nothing these birds do is slow!

In winter, when fewer flowers are blooming, hummingbirds will add insects to their diet to get the extra weight and nutrients they need for their spring migration. It takes them about two weeks to get to the southern U.S. from the Caribbean, then another 6 weeks for those that go all the way up to Canada. They start their journey north in early May and begin their return trip in July. If they survive, that is...

One of the craziest things about hummingbirds is that, because they're so small, they have a wide range of predators... including insects! Praying mantises have actually been observed hunting hummingbirds at flowers and artificial feeders. The mantis will lie in wait, perfectly still, and when the time is right, they deliver a powerful strike. That immobilizes the bird, and the mantis starts to eat it alive. Yikes! Hummingbird predators also include spiders, dragonflies, frogs, and even other birds (like the loggerhead shrike we talked about earlier!). It's a little sad – but mostly fascinating – the way nature works.

Another fascinating fact? Hummingbirds can see ultraviolet light! Let's compare... we humans have three color-sensing cones in our eyes that allow us to see the colors of the rainbow. Hummingbirds have 4; they can see the colors of the rainbow, plus the UV spectrum that's invisible to humans. Their 4th cone allows them to see shorter wavelength light that humans can't see. Do an internet search for UV-green and UV-red images – you'll see the psychedelic world that's visible to a hummingbird!

CHAPTER S
Sea Turtle
(see tur·tl)

I love highlighting animals I've had personal experiences with! So, during my freshman year of college, I actually took a class specifically about sea turtle biology. One weekend, our instructor arranged for us to go to Topsail Beach in North Carolina to visit the Karen Beasley Sea Turtle Rescue and Rehabilitation Center. We learned all about how the biologists rehabilitated injured sea turtles before releasing them back into the wild. We also spent a couple of days monitoring a nest that the biologists thought would be hatching that weekend.

We spent the night on the beach both Friday and Saturday waiting in anticipation, but we ended up being too early. The nest hatched a couple of days after we left. That said, a different nest farther down the beach *did* hatch while we were there. We were able to see it before we left and conduct a post-hatch nest analysis. That involves digging up the sand until the hatched eggs are exposed, then counting the eggs to see how many turtles hatched; and also counting how many intact eggs *didn't* hatch (insert sad face here). Unless you actually see a mama sea turtle lay her eggs from start to finish, this is the only way to truly know how many eggs she laid and how successful the hatch was.

Imagine the thrill when the first thing we saw as we dug up the nest was a baby sea turtle who'd been left behind! See, when baby sea turtles hatch, they're a couple of feet under the sand. They typically all hatch at the same time and then collectively flap their little flipper legs, which pushes the sand down – in turn moving the turtles up – until they reach the surface.

Every once in a while, a late hatcher is left behind. Without the help of the group, a single tiny sea turtle can't push through the packed sand to reach the surface. After we found our little turtle and marveled at it for a while, we placed it into a bucket so it could be checked out by the turtle vet before being released on the beach later that evening.

Besides being checked by the turtle doc, there was another reason we waited until the evening to release the hatchling. You see, sea turtle nests typically hatch at night. Hatching under cover of darkness reduces the chances that hatchlings will be preyed upon by birds and other predators as they emerge from the nest. The hatchlings then follow the reflection of the moon off the water to find their way to the ocean.

Have you ever been to a beach and seen signs instructing beachfront residents to turn their lights off at night? That's because artificial lighting can disorient hatchlings as to the direction of the ocean, making them travel up the beach into the dunes instead of down into the water. Those turtles have a smaller chance of survival… the longer they're on the sand, the more likely they'll end up being some animal's dinner.

Another important thing to tell you is that, when we returned to the beach to release our hatching, we let it go at its nest site rather than putting it directly into the water. See, adult sea turtles always return to the beach they hatched from to lay their own eggs. Turtle biologists think a baby turtle's short trek to the ocean after hatching is crucial to it being able to find its beach again when it's an adult.

No one is exactly sure how sea turtles accomplish this, but it's thought that, as they're scurrying to the sea after emerging from the nest, they're able to tune into the magnetic pulls of the Earth's north and south poles to get their bearings. They remember their beach's "frequency," so to speak, throughout their juvenile years and into maturity, and are able to use that early memory to locate their beach's region when it comes time to lay eggs. Is that not insane?!

Before we move on, a note about Karen Beasley… she was a woman who loved sea turtles and was committed to their conservation. Throughout her childhood, she monitored sea turtle nests all along Topsail Island with her family. After she passed away, her family fulfilled her request to use the money from her estate to start the Topsail Turtle Project, which created a formal monitoring program for Topsail's sea turtles. They eventually opened the turtle rehabilitation center on the island, fulfilling her final wish and keeping her legacy alive. That story absolutely fills my heart! The center was so successful that it grew and grew and has since moved to a larger facility about a mile inland in Surf City, North Carolina. I have so much gratitude for Karen's passion and dedication – she is an example for all of us!

Now, back to the hatchlings. After making their way to the ocean, baby sea turtles head for shelter. In the Atlantic Ocean, many are found in the Sargasso Sea, an area in the Atlantic where four ocean currents converge to form a gyre. Within this gyre are tons of floating seaweed called sargassum. The turtles forage there and also use it for cover to hide from predators until they're big enough to swim freely without being disturbed.

Ocean gyres also exist in many other places of the world, and sea turtles can be found within most. But before the babies make it there, they will face even more predators, mostly larger fish, birds, sharks, and whales. On average, less than one sea turtle hatchling out of 1,000 makes it to adulthood. And it takes 20-30 years for sea turtles to reach maturity, so if they're not protected, sea turtle populations could be at risk as fewer and fewer babies make it to adulthood to reproduce.

That said, sea turtles face much more dangerous consequences from humans than they do from their natural predators. For generations, both sea turtles and their eggs have been poached for food and shell ornaments. Sea turtles are also caught as bycatch in fishermen's nets and usually drown (bycatch is accidental capture of unwanted species by fishermen).

Six species of sea turtles are classified as vulnerable, endangered, or critically endangered: the leatherback (vulnerable), the loggerhead (vulnerable), the olive ridley (vulnerable), the green (endangered), the

Kemp's ridley (critically endangered), and the hawksbill (critically endangered). The flatback is listed as "data deficient."

Sad, right? That's why we all need to do our part to keep them safe. As for what we as individuals can do… if you're ever visiting the beach during the summer months, remember these three words: clean, dark, flat. Clean up your beach toys and trash, making sure to take everything with you when you leave; keep beach-facing lights turned off during nesting season; flatten any sand castles you build and fill in any holes you dig before leaving the beach.

Though sea turtles lay nests on the sand, they're built for living in the ocean. It's hard enough for them to drag their large, lumbering bodies across the beach to find a nesting spot and even harder when they encounter obstacles. Sea turtles can easily fall into large holes and get rolled onto their backs, with no way to right themselves.

As well as keeping beaches dark and free of debris and keeping plastic and other trash out of the ocean, we can advocate for sustainable fishing, not just for sea turtles but for all marine animals. Where turtles are concerned, devices like TEDs were introduced in commercial fishing several years ago. TED stands for Turtle Excluder Device and is a modified fishing net that has an escape hatch for turtles that get caught. Cool, right?

The way turtles migrate, they're more likely to get caught by commercial fishermen than more sedentary species. Did you know they swim thousands of miles over the course of their lives? Over 10,000 miles in many cases! They also have the ability to dive deeper than any other reptile, or even mammal. Yep, they can dive deeper than the deepest-diving whale! Sea turtles swim between foraging areas and nesting beaches all year, every year. Since many can live to be 100 years old, they cover a lot of ground – I mean, water – in their lifetimes!

Think about that... they leave the nest the day they hatch, head straight to the ocean, travel thousands of miles over the next couple decades, then return back to the same beach they were born to lay their own eggs. Or, at least, *very* near to the same beach. Could you do that? Leave home the day you were born, travel all over the world for several years (without a map to guide you), and then return to your family's doorstep 20-30 years later? No way! Sea turtles are awesome!

CHAPTER T
Termite and Tamandua

Termite

(**tur**·mite)

There are 2,800-ish species of termites alive today. These insects are so well-adapted that they live in all weather conditions and are found on every continent except Antarctica! Depending on the species, adults can be 1/10 to half-an-inch long, but the queen can be up to 10 times longer, reaching up to 4 inches!

Most termites are blind, so they communicate with each other through hormones and vibrations. Termites release pheromones, which are hormones that communicate various information to other termites, perhaps that a threat is nearby or that a certain task needs to be done for the colony. The vibrations, though? Termites usually reserve those for when a predator comes. They make those vibrations by pounding their heads against the tunnel walls in the colony. Yep, termites are basically head-bangers at a heavy metal concert!

The largest termite colony ever counted had 3 million termites, though it's believed that colonies can consist of up to 5 million individuals. Just like ants and bees, specific termites have specific roles in the colony: workers, soldiers, and queens. A colony is made up of mostly workers while it's being built and is growing. Once it's large and established, the queen will make more soldiers to defend the colony. In some species, she does this by… wait for it… by pooping on them! No kidding. The queen actually feeds her offspring pheromone-laced poop, and the hormones she injects into the poop determine what role the offspring will take on when they mature. How fascinating… and gross!

The queen will exclusively mate with 5-10 males – her kings – and produce sterile offspring for the first five years or so. After that time, her colony is pretty well established; after all, she lays 30,000 eggs a day. That's right, 30,000 *each day*. That's over 10 million a year.

After those first five years, the queen will then start producing fertile offspring. These termites will leave the colony once they've reached maturity to find their own mate and start their own colony. Worker and soldier termites live between 1 and 2 years, but queens can live more than 25 years; some African species are believed to live more than 50 years. That's insanely long for an insect! In fact, termite queens are believed to be the longest-living insects in the world!

They're not just long-lived; their whole evolution has been long. While there is some question as to just how long – with many scientists estimating 250 million years – the consensus seems to be that termites have been around for *at least* 120 million years. We have definitive fossils from the Cretaceous period, which spanned a period of time between 66 and 145 million years ago. And get this… termites evolved from cockroaches!

Here's an insane fact: the weight of all the termites in the world is greater than the weight of all the humans in the world! If you piled up every termite on earth and put them on a scale, they'd weigh around 445 million tons, give or take a few tons. If you piled up every *human* on earth and put them on a scale, they'd weigh around 350 million tons, give or take a few tons. Yes, there are 100 million tons more termites on earth than people!

Okay, I want to give you an idea of just how many individual termites that is. So, we're going to do a little math. This math comes with a lot of caveats, so keep that in mind. First, most termites are worker termites; they make up 90-98% of a colony. For the sake of simplicity, we're going to focus on workers.

Worker termites typically weigh around 1 milligram. That's 0.000002 pounds. That means it takes about 500,000 termites to equal a pound. There are 2,000 pounds in a ton. That means it takes 1 billion termites to equal a ton. *That* means it takes 445,000,000,000,000,000 (445 quadrillion) termites to equal 445 million tons! That's right, 445 plus 15 zeros! It's almost beyond imagination…

And the fun facts just keep coming… termites never sleep. Not once in their entire lives. They work on the colony for 24 hours a day,

7 days a week, until they die. Most termite colonies live inside wood, either in the ground or in building structures. Some dig tunnels underground that lead to a wood food source. Others build mounds and eat nearby leaf litter, rotting wood, and fungus.

I came across some termite mounds while traveling through Australia. They were amazing – and huge – structures. Termite mounds can easily be over 16 feet tall – that's three of me! The tallest termite mound ever measured was 42 feet tall! They've also been measured sprawling up to 100 feet wide.

A single termite colony can consume 1,000 pounds of wood, their primary food source, per year. In order to have the energy required to work 24/7 forever, termites need to eat *a lot*. Well, most of them do. Soldier termites, which usually make up about 2% of the colony, are unable to eat. Their mandibles – aka. jaws – are big enough to bite predators and colony intruders but far *too* big to bite into small plant material. So, soldiers rely on workers to regurgitate food into their mouths. Yep, soldiers eat worker vomit!

Tamandua
(tuh·**man**·doo·uh)

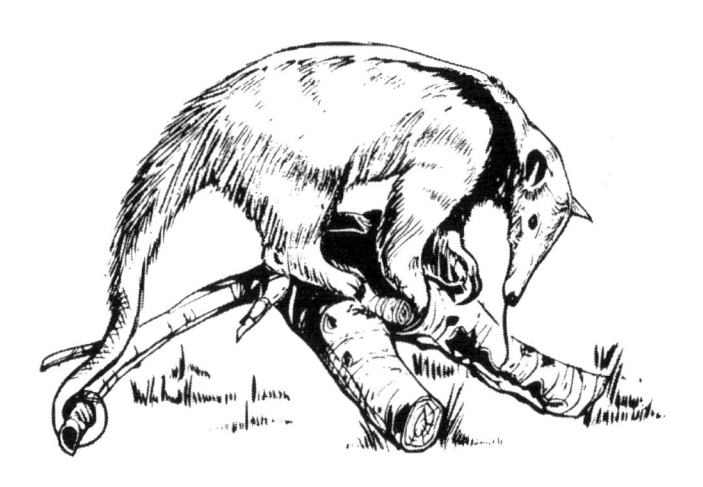

I had the privilege of caring for a tamandua at the aquarium where I interned in college. Just like our electric eel, we acquired it to go in our Amazon exhibit – they're found throughout South America in all kinds of habitats but are most abundant in areas adjacent to rivers. With its slender head, long snout, and extra-long, skinny tongue, it was the funniest animal I've ever had in my care.

Another name for the tamandua is the lesser anteater. You know, those animals with long tongues that slurp up insects? In fact, the tamandua's taxonomical suborder, *Vermilingua*, is a word that literally means "worm tongue!"

THE BONKERS BIOLOGIST

Tamanduas' bodies typically range in size from 1.5-2.5 feet long, but their tails add another 1-2 feet to their length. Their tails are so long that they can twist them up to their heads, with several inches to spare. They can often be observed using their tails as pillows when they're resting! Tamandua tails are prehensile, and the underside, which is used to grip tree branches for added stability, is hairless to provide a more secure grip. Sometimes, when I was entering my tamandua's exhibit, he'd balance on his back legs and tail with his front arms outstretched to grab hold of whatever I was bringing him. Man, was it cute!

Tamanduas top out at around 10 pounds. The one I cared for was about 2 feet long and 8 pounds. Their coloring ranges from dark brown to light blonde; some are darker on their shoulders and back, while some are a uniform color all over. Mine was a uniform medium brown with dense fur.

A tamandua's fur is for more than warmth; it actually protects the tamandua from ants and other biting bugs while it's feasting on a large colony. The insects can't penetrate its dense fur to bite or sting its skin. That said, its fur is shorter and thinner around its eyes, ears, and snout, so it is susceptible to more bites there. So, tamanduas only feed for short periods, so the colony it's feeding on doesn't have time to mount a massive defense. It moves on to another food source to let the prior one calm down before coming back for more. They can eat up to 9,000 insects a day!

Tamanduas can only eat small prey since their mouths don't open wider than ¼ inch in diameter – that's the diameter of a pencil. So, they're limited to leaner insects but can also eat honey, sap, or the insides of soft fruits that they tear apart with their claws. While they swallow their food whole, they have "toothed" stomachs that grind

A-Z ANIMAL FACTS FOR KIDS

their prey. Okay, their stomachs don't have *actual* teeth, but a hard gizzard that's good at breaking up food.

As part of my tamandua's enrichment activities, I'd take a clear water bottle, fill it with ants, then top it off loosely with dirt. Then I'd poke a hole in the cap, screw it back on the bottle, and toss it in the enclosure. The anteater would grab hold of the bottle and sniff around it until he found the hole in the cap. He'd thread his tongue through and start lapping up the ants at the bottom. It was fascinating – not to mention amusing – to see his long, skinny tongue flicking down the sides of the clear bottle! A tamandua's tongue is up to 16 inches long with sticky saliva and small barbs that get prey stuck to it like Velcro, similar to the echidna we discussed in Chapter E! Also, like echidnas, tamanduas have one of the lowest body temperatures of any mammal – just 91 degrees.

Tamanduas have large claws like their other anteater cousins, so they can break apart logs and dig through the soil in pursuit of the insects they eat. These claws are found on all four feet, and the back feet are used for gripping while the front are busy digging for food. The front feet have four toes, and there's an extra-large claw on the third toe, sort of like a velociraptor, that's especially useful for gripping, digging, and defense. What it's not useful for, though, is walking. When on the ground, tamanduas walk on the outsides of their feet, so their curved claws don't stab their soles! Because of their hazardous claws, they spend a lot of time in trees.

What if I told you that tamanduas and skunks have something in common? Can you guess what it might be? Here's a hint: think about the trait that skunks are most well-known for... yep, their stink! When they feel threatened, skunks release a foul-smelling odor for defense. No, it isn't gas... they don't fart on their threats! They release the odor from two separate, specialized glands on either side of their butts.

Well, Tamanduas have something similar. First, I can tell you from experience that tamanduas smell bad even when they don't mean to. They have really bad body odor, because all those insects they eat get really potent as they move through the digestive tract. And their poop? Don't get me started! It's smellier than my kid's blowouts when he was a baby! But their stink glands are a whole other ball game. Researchers think their spray is four times stronger than that of a skunk!

CHAPTER U
Uakari and Unau

Uakari

(wuh·**kaa**·ree)

Uakaris are small primates in the monkey family *Pitheciidae* that are found in the Amazon rainforest. They grow to between 14 and 23 inches long and weigh between 5 to 8

pounds. Scientists have divided monkeys into two groups: New World monkeys and Old World monkeys. They're separated this way primarily because of their physical features.

Simply put, Old World monkeys have noses with nostrils that open downward, like humans, and New World monkeys have flat noses with nostrils facing outward. Old World monkeys have opposable thumbs, like us, while New World monkeys don't. Old World monkeys are found throughout Africa and Asia, while New World monkeys are primarily found in South American rainforest habitats.

Old World monkeys may or may not have tails, but they aren't prehensile; as in, they can't grasp things like branches. New World monkeys not only have tails, but they *are* prehensile. They're so strong, they can usually hold the entire weight of the monkey. That's why several observations have been made of New World monkeys hanging from trees solely by their tails!

Here's where the confusing part comes in… Uakaris are New World monkeys, but they don't have long tails like their cousins. Uakaris have very short tails compared to other New World monkeys. So short, in fact, that they can't really be used for grasping things. Uakaris definitely can't use their tails to hang from trees. That said, they're still useful for balance. Uakaris can leap an astonishing 65 feet between trees due to some insanely powerful legs. So, they need all the stability they can get! That's where those tiny tails come in handy – uakari's stick them out for extra balance, kind of like we humans outstretch our arms for extra balance.

Uakaris *do* have flat noses; so, despite their opposable thumbs and odd tail, they're still classified as New World monkeys! There are four species of uakari that all look fairly similar except for fur color. From the neck down, their uniform coloring can be red, brown, black, or white. From the neck up, though? That's where things get amusing!

Uakaris have hairless, bright-red faces. They literally look like a person with a terrible sunburn! The long hair that covers their bodies goes all the way up to the crowns of their heads, then stops to reveal a bald, greyish-brown top-of-the-head. Starting at their foreheads, just below a very distinct widow's peak, is where their coloring changes to a deep pinkish-red. Essentially, their faces look like bright pink hearts! It really is funny!

Unau
(**oo**·naw)

Ever heard of an unau? I bet you have! But maybe by a different name... how about a sloth? Ever heard of that? That's what an unau is! The unau is also known as Linnaeus's two-toed sloth or the southern two-toed sloth. This species is found in South America, north of the Amazon River.

Unaus are one of six species of sloth, all found in Central and South America. There are both two-toed and three-toed sloths, and most are 2-2.5 feet long and weigh between 8 and 17 pounds once fully grown. While today's sloths are about the size of small- to medium-sized dogs, recent research published in 2019 confirms that they all evolved from ancient ground sloths. These included the *Mylodon*, which was the size of a horse, and the *Megatherium*, which was the size of... wait for it... an elephant!

Sloths are very, very slow-moving animals, only covering about 100 feet a day. Scientists think this is because of their slow metabolism and a diet that's not very high in energizing nutrients. Three-toed sloths are herbivores and only eat plants, while two-toed sloths like the unau are omnivores. In addition to plants, they'll also eat lizards and insects if they can catch them. But here's the kicker... it takes them between 2 weeks and one month to digest a single meal. One month!

Sloths have specialized multi-chambered stomachs, like cows, that are pretty much always full (their digestion is just as slow as everything else they do). They hold almost a third of their weight inside their stomachs! That also means that they can lose a third of their body weight when they poop!

Their stomachs are surrounded by a lot of ribs, especially the unau and other two-fingered species. Humans have 12 sets of ribs, but two-toed sloths have nearly double that! Twenty-three pairs, or a grand total of 46 individual ribs! That's the most of any mammal on earth.

Despite sloths taking forever to do pretty much anything, they do have some pretty amazing defense mechanisms. They actually wouldn't have one of these defense mechanisms if they weren't so slow. It's their great camouflage. Their fur is a brownish color that blends into tree trunks and branches. But, because they're so sedentary – sleeping up to 20 hours a day and then moving super slow when they're awake – it gives algae the opportunity to grow in their fur. So, most sloths appear to have a green tint, further camouflaging them in the tree canopy!

Sloths have really long arms with sharp claws. They don't need to use them too often, but when threatened, they can take a swipe at a predator, leaving a nasty gash. They mostly stay up in the trees to avoid threats altogether, though. Their tendons have unusual locking mechanisms that allow them to hang from branches by their arms and legs without using a single drop of their limited energy. They simply find a comfy position, then lock the tendons in their hands or feet in place, taking 100% of their weight off their arm muscles!

So, while a human's arms would give out after hanging like that for just a little a while, sloths can sleep in a hanging position all day. If they didn't need to move around to forage, a sloth could conceivably hang from its arms or legs forever. In fact, if a sloth dies in a hanging position, they sometimes won't fall out of the tree – once their tendons are locked in place, they remain that way even after death! Can you imagine ever coming across a dead sloth hanging from a tree? Crazy!

Even without their special tendon-locking abilities, sloths are super strong for their size. When babies are born, they already have the ability to not only hang from a tree with one arm but also do a pull-up with one arm! They can't spend all day doing pull-ups, though. While strong for their size, remember their limited energy? They have to conserve what little they have.

Sloths also eat poisonous leaves. You know poison ivy, right? The stuff that gives us an icky rash when we touch it? Sloths eat it and other toxic leaves on purpose. They have specialized multi-chambered stomachs that can digest the rubbery rainforest leaves they eat, as well as process the toxins inside them. The toxins stay in their bodies, though, giving them the ability to poison predators that bite them. Sometimes, if a large snake or bird-of-prey tries to eat a sloth, they come into contact with the poison in the sloth's bloodstream, which makes their throats swell so much that they can't breathe. So, in a way, sloths can poison their predators to death!

Okay, one last fact involving death, and then we'll move on to something happier… if it gets too cold out, a sloth can die of starvation. Wait, huh? What does one have to do with the other? Well, in order to further conserve their limited energy, sloths don't have the ability to self-regulate their body temperature like other mammals. That's why they're often observed sunning themselves in the trees.

But if it gets too cold out, the digestive bacteria in a sloth's stomach will die. Once that happens, they can no longer digest food. And once that happens, their bodies can no longer absorb the nutrients they need to survive. So, even if they've eaten and are completely full, they can technically die of starvation. Yes, an animal can die of starvation with a stomach full of food. Insane!

Okay, enough death talk. Remember the sloths' algae-covered fur? Well, it isn't just home to algae. Thousands of insects and bugs representing hundreds of species can be found living on a single sloth. Their fur supports whole unique ecosystems and, while I wouldn't want a bunch of bugs on my own body, it's fascinating that sloths can support so many insects at different phases of their life cycles. In fact, while bug infestations are usually a sign of an ailing animal, an insect-covered sloth is a healthy sloth! There are even six moth species that depend on sloths for their survival. Affectionately called "sloth moths," they simply wouldn't exist if they didn't have sloth fur to live in and sloth poop to lay their eggs in (gross, I know, but also kind of cool!).

As cool as it is, though, the sloth's slow-moving nature makes it an easy target for other animals looking to chow down on some tasty insects. Birds, and even Capuchin monkeys, have been observed picking moths out of a sloth's fur to eat!

CHAPTER V
Vampire Squid and Virgin Islands Dwarf Gecko

Vampire Squid
(**vam**·pai·ur skwid)

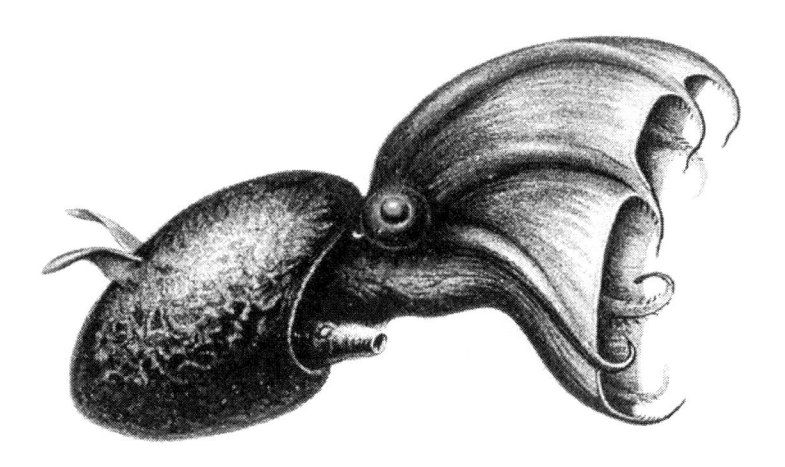

Vampire squid are neither vampires (duh!) nor squid. They *are* cephalopods, like squid and octopi, but they branched off from common ancestors and eventually evolved into their own classification, now a distant relative of squid and octopi.

Vampire squid get their name because they have eight arms with two additional tentacles, like squid; plus, many have black bodies with red eyes and spine-like protrusions coming from the underside of their webbed arms. Someone must've thought all those characteristics resembled Dracula. Hence, vampire squid!

Vampire squid live so deep in the ocean that they can survive in pitch black waters that are nearly freezing cold – temperatures that would cause hypothermia in humans in just minutes. And get this… the Guinness Book of World Records recognizes 1,090 feet as the deepest depth any human has ever scuba-dived. That's just one third of the depth that vampire squid can live – they're found as deep as 3,000 feet!

One of the coolest things about the vampire squid is that it has the largest eyes in the entire animal kingdom – relative to size, of course. Adults measure about a foot long with eyes up to an inch in diameter. Let's compare… if a human had eyes as big as a vampire squid, then, relatively speaking, a person who stood 6 feet tall would have eyes 6 inches in diameter!

Female vampire squid are super committed mamas. They carry their eggs for about 13 months before laying them, and they don't eat a single thing the whole time! Some are so exhausted by the time they lay their eggs that they die of exhaustion. Yep, they're pregnant for over a year and fast throughout, only to die when they finally reach the finish line – no, thank you!

Do you know that squid and octopi release dark ink when they're threatened? Not the vampire squid. Ink wouldn't do any good at deterring predators in the dark depths they live in. So instead, they shoot out bioluminescent mucus (that means their mucus emits light)! It sparkles for about 10 minutes, giving the vampire squid plenty of time to escape from a confused predator. Bioluminescent snot… can you imagine blowing that into a tissue?!

The tips of each vampire squid arm are also bioluminescent. Scientists think this is primarily used to identify other vampire squid. It is believed that vampire squid are solitary and don't likely run into their kind very often. So, at the dark depths they live in, it would be hard to find a mate without help. Hence, bioluminescent arms to locate each other.

Vampire squid are like the vultures of the ocean – they don't kill their prey but rather scavenge dead animals. Since they live so deep – remember, up to 3,000 feet! – this usually consists of things that fall from higher up in the water column. Vampire squid eat dead plankton and other small creatures, marine "snow" (which is basically any organic matter that sinks from higher up), and, most amusingly… poop! Yep, when other animals poop, the vampire squid is waiting below to gobble it up – unless it's an animal that poops out floaters!

THE BONKERS BIOLOGIST

Virgin Islands Dwarf Gecko
(**vur**·jn **ai**·luhndz **dworf geh**·kow)

About the size of a U.S. dime, the Virgin Islands dwarf gecko is one of the smallest reptiles in the world today! Its body is less than ¾ an inch fully grown. It weighs in at five-thousandths of an ounce, equivalent to 3 ten-thousandths of a pound! Most insects that size weigh more than the dwarf gecko!

The title for smallest reptile goes to nano-chameleons, which can rest on the tip of an adult's pinky finger with room to spare. It's thought that males top out at just over half an inch, and females just *under* ¾ inch. So, the Virgin Islands (V.I.) dwarf gecko isn't much bigger. Can you imagine holding a lizard on the tip of your finger? I cannot *wait* to have that chance.

As the name implies, V.I. dwarf geckos are only found in the Virgin Islands. Not only is their range restricted to the Virgin Islands, most scientists think they're endemic to only the British Virgin Islands of Virgin Gorda, Tortola, and Moskito. As of 2021, they hadn't been observed on any other island.

Since they're so small, scientists haven't been able to conduct robust surveys in the wild, and therefore they don't know much about the V.I. dwarf gecko's life history. What scientists *do* know is that they lose water much faster than other geckos – about 70 % faster, according to researchers – so they spend most of their time and energy retaining as much moisture as possible. Even though they lose moisture so quickly, a necessary thing for a gecko's survival, V.I. dwarf geckos don't live in the coastal areas of the islands. They live in arid environments, mostly dry forests up in the hills (psssst… arid means an area that doesn't get a lot of rain).

Arid environments don't sound like the best choice given V.I. dwarf geckos enhanced moisture needs, but they make it work. The dwarf geckos keep from drying out by taking cover in wet, shady places – they lay low under logs and moist ground vegetation for most of the day.

Just like other geckos, V.I. dwarf geckos don't have eyelids. So, instead of blinking to keep their eyeballs moist, like other animals, geckos lick their eyeballs! And since V.I. dwarf geckos lose moisture so much quicker than their relatives, I wonder if this means they lick their eyeballs more often, too. Inquiring minds want to know (and this inquiring mind couldn't find a definitive answer)!

Like most geckos, V.I. dwarf geckos use a sticky tongue to snap up insects, but they're so small that they're limited in the prey they can eat. Scientists haven't been able to study the diet of V.I. dwarf geckos specifically, but plenty of studies have been done on other tiny geckos in the same *Sphaerodactylus* genus. Assuming they behave similarly, it's reasonable to conclude that springtails are a main staple in the dwarf's diet since these bugs are typically less than ¼ inch long.

Though it hasn't been studied in V.I. dwarf geckos, scientists also believe they share the trait of fat storage with their other gecko relatives. Most geckos have the ability to store fat in their tails to use in times of food scarcity!

Further, if they share additional life cycle traits with other *Sphaerodactylus* species, V.I. dwarf geckos likely mature between 1.5-2 years, with a maximum life span of 4-6 years. Their full scientific name is *Sphaerodactylus parthenopion*, and *Sphaerodactylus* means "round finger." The name makes total sense because most geckos do actually have rounded toes on their feet.

And did you know those feet are magnetic? Okay, okay... they're not actual magnets, but get this: electrons on geckos' feet interact with electrons on the surfaces they climb on to create an electromagnetic force. That's what makes them stick to smooth surfaces like walls, ceilings, and windows so effortlessly!

V.I. dwarf geckos also have an added feature for sticking – an adhesive scale on each of its 4 feet!

But those electrons? They're found in hairs on the bottoms of the gecko's feet called setae. And get this: those hairs have hair! Yep, each seta has hundreds of super tiny additional hairs sprouting from it. They contribute further to the gecko's sticking power. On top of that, geckos can point the setae at different angles to increase or decrease their sticking power. That's exactly what they do when they want to jump from one smooth surface to another. They can literally turn their sticking power on and off!

CHAPTER W
Whelk and Whale Shark

Whelk
(welk)

Whelks are essentially snails that live in the sea. And I learned something very amusing about whelks – but also a little bit sad – when I was working on my master's degree in wildlife biology: in the early 2000s, researchers found something strange going on with female whelks in Australian harbors... they were growing male reproductive organs out of their heads!

That's the amusing part. Here comes the sad part... it was because of us. Humans have done some real damage to fish and wildlife populations since the industrial revolution. In this case, the culprit was manmade TBT. TBT is short for tributyltin, a compound in antifouling paint.

Boat and ship owners loved antifouling paint because it repelled whelks, snails, barnacles, algae, and other sea life from attaching to the underside of boats. When a bunch of marine life settles on boat hulls, it creates more drag that makes ships move more slowly and, therefore, cost more money to run.

The problem is, TBT is an endocrine disruptor. This means that it interferes with an animal's hormones, including reproductive hormones. This unfortunate phenomenon has been known for a while, but it was still a shock when researchers saw lady whelks with male reproductive parts *growing out of their heads*! Even crazier (and sadder)... the presence of these male parts interfered with the female's ability to shed her eggs. Instead, her collection of eggs would continue to grow and grow inside her abdomen until she exploded!

Luckily, TBT was banned in the late 2000s, and Australia's whelks are recovering nicely. Instances of females with male parts have declined significantly, and researchers think this trend will continue. Thank goodness we recognized the problem and intervened in time!

Let's stay on the subject of marine animals, but without the heinous side effects of TBT...

Whale Shark
(wayl shaark)

The mighty whale shark is the largest fish in the ocean, weighing in at about 20,000 pounds with an average length of 40 feet, though individuals over 60 feet long have been reported! To put that into perspective, the great whites you typically see on television average 15-ish feet in length.

THE BONKERS BIOLOGIST

Whale sharks live longer than humans, with average life spans between 80 and 130 years. Some long-living individuals can even top 150 years old! They don't even reach mating age until about 25.

The most interesting thing about whale sharks? They don't use their teeth! Wait... aren't sharks synonymous with teeth? Sure, and whale sharks have plenty – around 350 *rows*. But they are one of 3 shark species that are filter feeders. Whale sharks swim through the water with their gigantic mouths open, taking in tiny krill and plankton as they go. Then, they force the water out through their gills and swallow the food that's left.

Crazy to think that the largest fish in the ocean eats the smallest creatures in the ocean, but it's not the only one. Blue whales, a species of baleen whale that can top 100 feet long and 400,000 pounds, are both the largest mammals *and* the largest animals in the world. They're also filter feeders that eat the tiniest prey, including krill, plankton, algae, and small fish.

That's where whale sharks get their confusing name. They are technically sharks but filter feed like baleen whales.

After the Jaws movies came out in the 1980s, sharks got a pretty bad rap. And, yes, while there are instances of shark attacks every year, the number is very minuscule. And it's never the result of a whale shark encounter because they are absolutely harmless to humans – in fact, they're harmless to all animals except their prey. A person can actually swim with a whale shark without any threat. (Side note: I'm not advocating swimming with whale sharks or interacting with any wild animals up close. If you ever encounter one, give them a wide berth and observe from a distance, please!)

That said, I'd certainly get a good fright if I ever saw a whale shark swimming toward me with an open mouth – they open to about 5 feet wide! And get this… there's at least one report of a diver getting caught in a whale shark's mouth (though I can't confirm that via any reputable source – I can only find the incident described in an online dive report). According to the report, the shark simply spit the diver out!

But even if whale sharks *were* aggressive toward humans, we'd be safe from being swallowed. Though they have gigantic mouths, their esophagus is tiny in comparison. Put your thumb and index finger together to form a circle. A whale shark's esophagus is actually smaller than that circle – it's about the diameter of a U.S. quarter! Weird, right?!

Because of the minuscule size of their prey, whale sharks don't need a large throat. A huge *mouth* is necessary for filter feeders since they need to take in a lot of water to get a good volume of food. But a huge *throat* isn't necessary since the plankton and krill they filter out to eat are so small. Speaking of filter-feeding, whale sharks can take in and filter out up to 150,000 gallons of water an hour when feeding. What?!

CHAPTER X
Xingu River Ray and Xanthippe's Shrew

Xingu River Ray

(shing·**goo** **ri**·vr ray)

Y ou're probably well aware of ocean-dwelling stingrays. But did you know there are freshwater rays, too? Around 35 species of them! And since we're on Chapter "X," we'll be talking about the Xingu River ray.

THE BONKERS BIOLOGIST

The Xingu River is in Brazil and a tributary of the larger Amazon River. The Amazon and its tributaries are the only known rivers where the Xingu River ray lives. Scientists aren't sure of this ray's conservation status because there just isn't much information on the species. But if its plight is anything like its relatives, it could be suffering the effects of habitat degradation.

Huge swaths of land are constantly being razed for the logging business and to make room for plant and animal agriculture. Runoff from grazing livestock – specifically, their poop! – seriously pollutes the water and threatens the health of its inhabitants, including river rays. Since the range of the Xingu River ray is so limited – to only the Xingu River and two of its tributaries – unhealthy water quality could pose a serious problem to their long-term survival if it's not corrected.

One unique tidbit about the Xingu River ray is that it gives birth to a lot of babies in comparison to other stingray species. The average stingray has around four pups per litter, but the Xingu River ray averages eight pups, and sometimes as many as a dozen! Mama Xingu rays keep their eggs inside their bodies, where they eventually hatch. That means live pups emerge when they're born, looking like teeny tiny versions of their parents.

Mature Xingu rays are about 2 feet long, including their tail, and their bodies are 12-16 inches in diameter. Not much information is readily available on exactly how large pups are when they're born, but if other rays are any indication, they're probably a couple of inches in diameter. I've never seen any species of river ray pups in person, but I have seen newborn ocean rays with bodies just a couple inches across… and, my goodness, are they cute!

Xingu River rays have bodies that are pretty much round, which is why they're called discs, unlike other rays that have somewhat pointed snouts and "wings." Their typical coloring is black with white-ish spots, which help them camouflage on the rocky bottom of the river. The white spots on top of the disc also resemble sun glints, which is when light penetrates the water and reflects off the riverbed, further helping them blend into their environment. They can also burrow just under the top layer of the riverbed to hide from predators, and, as a last resort, they have a stinging barb on their tail to defend themselves.

Even though they don't actively try to sting other animals, they're responsible for a lot of injuries to people. This is because they're typically very docile, usually just sitting still on the river bottom, so they're pretty easy to step on! In fact, if you combine the stings of all Amazonian river rays, they collectively injure more humans than any other animal species in the Amazon!

Xanthippe's Shrew
(zan·**thi**·peez **shroo**)

The word shrew is a noun that has two definitions: 1 – an insectivore mammal with small eyes and a pointed snout that looks like a mouse; 2 – an argumentative, ill-tempered woman.

Have you ever heard of the philosopher Socrates? His wife was named Xanthippe, and she was often referred to as a shrew because of her fierce, confrontational nature. And, when Shakespeare wrote *The Taming of the Shrew,* he meant the woman, not the mammal!!

Put the name and the noun together, and you have the Xanthippe's shrew! It kind of makes sense since shrews are solitary animals that can become pretty aggressive when other shrews, animals, or even people enter their territories.

Xanthippe's shrews belong to the sub-family *Crocidurinae* that includes white-toothed shrews. Other shrews have reddish teeth. But what all shrew teeth have in common is that those bad boys are sharp! Shrews have an insanely high metabolism and must eat every couple of hours to maintain their body weight. In fact, to do that, they have to eat three times their weight each day! That's how high their metabolism is.

Though shrews look like mice, they're a type of mole rather than a rodent. They're primarily found in Kenya and Tanzania, and they're found *a lot*. There are millions of shrews in the area because they reproduce so many times a year. Mama shrews have a gestation period of less than a month, and their babies become independent just three weeks after they're born. So, mama is free to reproduce, again and again, having up to 3 litters in a single year. Overachievers can squeeze in 4 litters a year!

So, imagine this cuteness… sometimes a mama shrew needs to move herself and her new babies to a different home before they're ready to go off on their own. How does she do it? She lines the babies up, face to butt, and each baby grabs hold of the baby in front of them; as in, they each take hold of a butt with their mouth. Mom heads up the line, and they all move single file to their new place before letting go… and then hopefully washing out their mouths!

When they arrive at their new home, it isn't one mom has built, but one she's found. In fact, shrews hardly ever build their own homes. Remember that high metabolism? They can't afford to expend all the energy that's required to dig burrows. Instead, they use abandoned ones usually dug by moles and other underground-dwelling species. Shrews definitely follow the old adage, "Work smarter, not harder!"

CHAPTER Y
Yeti Crab
(yeh·tee krab)

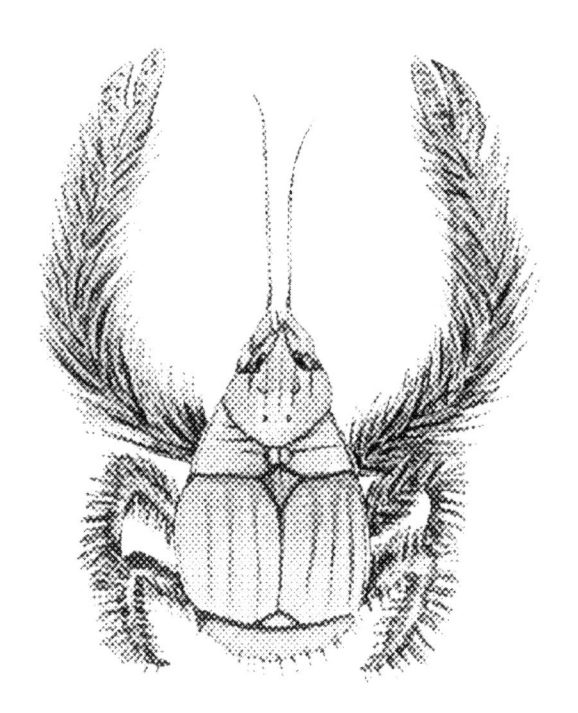

What do you need to know about the Yeti crab? I'll admit, this is a weird pick... because there isn't a lot of information about this recently discovered species! But it's so darn funny-looking, I needed to share!

THE BONKERS BIOLOGIST

A yeti crab looks exactly like its name implies: an abominable snowman with the body of a crab. This white-to-light-gray creature has eight legs sporting long, light yellow hair! You must Google it ASAP. I'll wait...

Yeti crabs were first discovered in 2005, not only as a new species but as a new taxonomical genus *and* new taxonomical family! (Remember the levels: kingdom, phylum, class, order, family, genus, species.) Why were they only recently discovered? Because they are bottom dwellers living on the deepest parts of the ocean floor near hydrothermal vents, places only reachable by specialized research submarines. The research team that discovered it was on an expedition studying deep-sea hydrothermal vents near Easter Island off the western coast of South America, where the Pacific-Antarctic ridge lies.

The Pacific-Antarctic ridge is where the edges of two tectonic plates meet – the Pacific Plate and the Antarctic Plate – in the south Pacific Ocean. There are a large number of hydrothermal vents along the boundary, and during this particular expedition, scientists were comparing species compositions between vents hundreds of miles apart. That's when the first yeti crab was spotted on the seafloor beside a crack in some lava rock where hot water was shooting out.

Yes, I said lava rock! Hydrothermal vents form near tectonic plate borders where lava rises up under the earth's crust. The crust cracks under all the pressure from the lava, and the lava heats the water that seeps into the cracks, which then flows back up into the ocean. This hot water has been infused with sulfur- and metal-rich minerals from the lava. Several specially adapted deep-sea animals get their nutrients from this water!

Since the yeti crab was only recently discovered, scientists still aren't sure exactly how it benefits from hydrothermal vents. But, during that initial research mission in 2005, the researchers observed one standing by a vent with its first two legs – which are clawed and much longer than the rest of its legs – outstretched and waving above the hot water that was spewing out of the cracks. The working theory is that it was trapping mineral-rich bacteria in the hair of its arms to eat later.

One thing scientists quickly determined is that yeti crabs are mostly blind, if not completely blind. So, the hairs on its arms could also be sensory receptors that help them navigate, locate food, and/or find mates.

Despite living near hydrothermal vents where the hot water can reach scalding temperatures over 700 degrees, it seems that yeti crabs still get cold. Researchers observed hundreds of crabs piled on top of each other, presumably for added warmth in the chilly deep sea. They were packed so tightly that 700 crabs were counted in a single square meter!

Temperature changes are extreme around hydrothermal vents. Because the ocean is so deep, the water at the bottom is cold, right around the freezing mark. But at the opening of a vent, the scalding water is so hot that it can kill many animals. So, organisms there have to find that sweet spot where they can take advantage of the warmth of the vent while not getting too close. On the other hand, if they get too far away, some could become too cold to survive. It's a balance that animals in these areas have mastered!

They've also adapted to life without the sun. The vast majority of animals and plants on earth rely on the sun for some part of their survival. Even we humans need the sun to get Vitamin D, an essential vitamin for absorbing calcium and developing strong bones. When we don't get enough, our doctors prescribe supplements.

Animals living near hydrothermal vents rely on the chemicals that flow from the vents to replace what most animals get from the sun. Whether its tiny bits of food, essential nutrients, or energy, they get it from the hot, sulfuric water that the magma has infused with all sorts of metals and minerals.

Back to yeti crabs… Get this: despite the name, yeti crabs aren't classified as "true crabs." Those folks who decide common names for new animal species need to stop making things so confusing, am I right? Yetis are more closely related to animals called "squat lobsters," a group that hermit crabs are also a member of. So, yeti crabs are distant relatives of hermit crabs… crazy, right? Hermit crabs live in warm, shallow tide pools as little as one inch deep, while yeti crabs live on the cold seafloor over a mile deep!

After yeti crabs were first discovered in 2005, researchers studying in other parts of the world knew to be on the lookout. And they found more! In 2006, 2010, 2011, and 2013, four additional yeti crab species were discovered, from Central America to the Indian Ocean to Antarctica!

I have to note here that I'm especially excited about the yeti crab's discovery because it was made by researchers from the Monterey Bay Aquarium in California. Several years ago, I contributed to the aquarium's Seafood Watch program as a contract writer, researching and compiling reports about various commercially fished species. They used my research, and that of others, to create consumer guides for the public so people can be informed about whether or not their seafood choices are sustainable and how to choose ones that are. It's something I'm super proud of!

CHAPTER Z
Zebra Shark and Zigzag Salamander

Zebra Shark
(**zee**·bruh **shaark**)

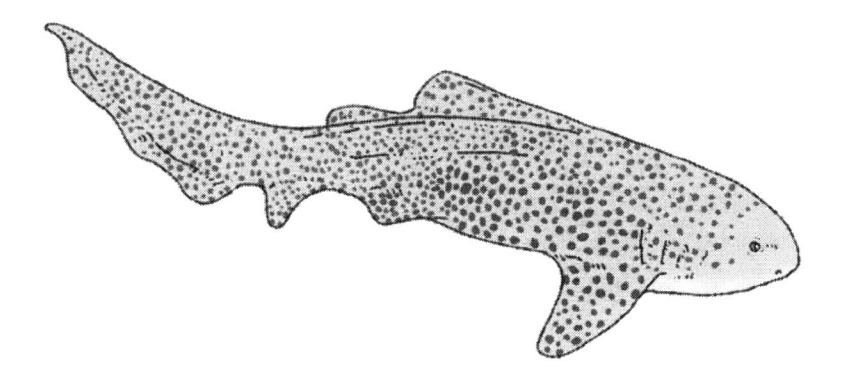

I magine what a zebra shark might look like… a shark with stripes, right? Wrong! Zebra sharks are spotted. Ugh – yet another animal name that makes no sense!

Actually, it makes sense for a little while because zebra sharks are born with brown skin and yellow stripes. As they age, the stripes become spots, and their coloring flips – their skin becomes lighter, and the spots become dark. Adults are sometimes mistaken for leopard sharks.

Besides the name, there are a few traits that make the zebra shark unique. They're slimmer and more flexible than other sharks, so they can forage for food in the nooks and crannies of coral reefs, pushing their bodies into small crevices to suck out their prey.

Yes, I said "suck out" their prey. If a tasty morsel is out of reach in the coral, the zebra shark can use its mouth like a powerful vacuum, sucking it right in! If you've ever seen divers hand-feeding sharks on television, there's a good chance they were zebra sharks, especially if they exhibited the food-sucking behavior.

Zebra sharks usually do their foraging at night after spending the day sleeping on the ocean floor. That doesn't stop them from seeing their prey, though. On the sides of their mouth, they have what look similar to whiskers. These are actually sensory organs called barbels that help the zebra shark locate food.

Perhaps the coolest thing about zebra sharks is how they mate and reproduce (yes, it's probably clear by now that I'm fascinated with animal reproduction!). First, let me mention that huge numbers of zebra sharks gather every summer off the northeast coast of Australia, a behavior that would typically indicate mating. But researchers haven't observed any mating during this zebra shark summit, and to this day, remain unsure of the purpose of the annual reunion.

When scientists *have* observed zebra shark mating, what they've seen is that the male will nip the female's tail to show his interest. Ouch!

After mating, female zebra sharks lay four eggs at a time over a period of several months. They can lay over 40 eggs in a season. A few years ago, a captive female at an aquarium in Australia laid four eggs, 3 of which hatched. It fascinated everyone, especially scientists… because she hadn't mated!

This zebra shark had reproduced by a process known as parthenogenesis, which is well known in the animal kingdom. But it's mostly observed in invertebrates and "simple" vertebrates, less so in more complex vertebrates. Parthenogenesis is basically asexual reproduction. In zebra sharks, a female can merge polar bodies – small cells that are a byproduct of egg creation – with fertile eggs. This genetic material takes the place of the male's contribution of chromosomes, resulting in viable offspring!

While this is super cool, it could indicate that mama zebra shark was stressed. Parthenogenesis is a strategy that some animals turn to in dire conditions, which they sense could be a threat to the survival of their family line. Since the captive shark at the aquarium was the only zebra shark in her exhibit, it's possible that her instinct to preserve her lineage took over, motivating her to reproduce on her own since she didn't have any indication that there were other zebra sharks out there to continue on. But that's just this wildlife biologist's opinion; I can't be sure!

THE BONKERS BIOLOGIST

Zigzag Salamander
(**zig**·zag **sa**·luh·man·dr)

There are three types of zigzag salamanders: the northern, southern, and Ozark zigzags, all found in moist forest habitats. Northern zigzags are found from Alabama up to Indiana; southern zigzags are found from Mississippi and Alabama up to southeast Virginia; and Ozark zigzags are found in the highland region of mountains and plateaus known as the Ozarks, primarily in Missouri.

Zigzags are on the small side, with body lengths just under 2 inches. Their bodies range in coloration from brown to gray. Most also have tiny spots that are lighter in color than the rest of their bodies, usually light gray to white... against a gray backdrop, these spots can give the salamanders a blue hue! The spots can be so tiny that they look like flecks of glitter against the salamander's moist, shiny bodies. It sort of makes them look metallic!

Zigzag salamanders get their name from the tell-tale stripe down the center of their backs. It can be yellow to orange to red and typically has jagged edges, though it's not visible on all of them. For those that don't have a very visible stripe, they can be identified by their red-orange skin coloration where each limb attaches to the body. What I'm saying is, they essentially have red armpits!

Zigzags are unique in that they don't have a larval stage. Most other salamanders do, hatching from an egg into juveniles that swim around like tadpoles before going through a metamorphosis to complete the transition to adulthood. Zigzags hatch out of their eggs already in their adult forms. They don't have to undergo any transformations to get to maturity; they just have to grow!

Zigzag salamanders can detach their tails to escape a predator that grabs them. And, like all salamanders, zigzags are also poisonous. They excrete poison from glands under the skin for defense from predators. I learned this the hard way when I was a child attending summer camp in the North Carolina mountains. I looooooved picking up any animal I came across, and salamanders were no different. But after playing with one for a bit and then releasing it, my hands were tingly for quite a while!

Did you know that zigzags, like the other salamanders in their taxonomic family, don't have lungs? Instead, they use what's called cutaneous respiration – which means oxygen is absorbed through their skin. Likewise, carbon dioxide is released through their skin too. Salamanders are so neat!

LEAVE A 1-CLICK REVIEW!

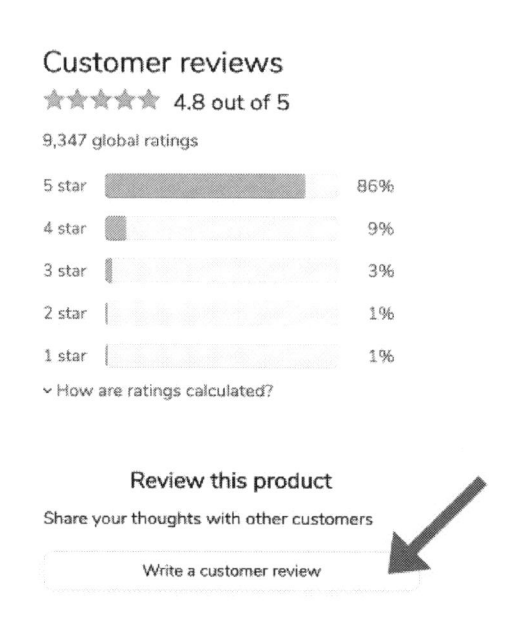

Customer reviews
★★★★★ 4.8 out of 5

9,347 global ratings

5 star	86%
4 star	9%
3 star	3%
2 star	1%
1 star	1%

How are ratings calculated?

Review this product

Share your thoughts with other customers

Write a customer review

I would be incredibly grateful if you could take just a few seconds to write a brief review on Amazon, even if it's just a couple sentences!

Scan the QR code or visit this link:

https://www.amazon.com/review/create-review/?asin=B0CPLLKNXK

IN CLOSING . . .

There you have it... the Boners Biologist's picks for the most amazingly amusing animals on the planet! Aren't they awesome?! Be on the lookout for more books about animals, nature, and the environment coming soon. Until then, I leave you with my favorite quote:

> "Look deep, deep into nature, and then you will understand everything better."
>
> ~**Albert Einstein,** *Einstein in Berlin*

DON'T FORGET YOUR FREE GIFT!

Amazingly Amusing Animal Trivia for Kids (and Their Adults!)

Test your knowledge of all the amazingly amusing facts you'll learn in this book! Quiz questions for older kids (and parents!), coloring pages for younger kids!

Scan the QR code or visit this link:

theBonkersBiologist.com/animal-trivia

REFERENCES

Armadillo

Armadillos. (2021). National Geographic. Retrieved 2021, from https://www.nationalgeographic.com/animals/mammals/facts/armadillos

Bradford, A. (2015, October 6). *Armadillo Facts.* LiveScience. Retrieved 2021, from https://www.livescience.com/52390-armadillos.html

Can Armadillos Roll Into A Ball? (2018). AnimalQuestions.Org. Retrieved 2021, from http://animalquestions.org/mammals/armadillos/can-armadillos-roll-into-a-ball/#:~:text=The%20only%20armadillo%20that%20is%20able%20to%20roll,tucking%20its%20head%20and%20legs%20into%20its%20shell

Arapaima

A-Z Animals. (2021, March 9). *Arapaima.* Retrieved 2021, from https://a-z-animals.com/animals/arapaima/

International Game Fish Association. (n.d.). *Arapaima (Arapaima gigas).* Retrieved 2021, from https://igfa.org/igfa-world-records-search/?search_type=CommonNameSummary&search_term_1=Arapaima

Majumdar, S. (2012, February). *Paiche — Iron Chef Ingredients.* Food Network. Retrieved 2021, from https://www.foodnetwork.com/fn-dish/shows/2012/02/what-is-paiche-iron-chef-ingredients

National Zoo. (n.d.). *Arapaima.* Smithsonian's National Zoo and Conservation Biology Institute. Retrieved 2021, from https://nationalzoo.si.edu/animals/arapaima

Sargent, B. (2015, August 8). Fly fisherman scores a giant record. *Florida Today.* https://www.floridatoday.com/story/sports/outdoors/bill-sargent/2015/08/08/fly-fisherman-scores-giant-record/31346857/

Bee

Abbott, C. (2020, August 4). Colony Collapse Toll is Highest in Four Years for U.S. Honeybees. *Successful Farming.* https://www.agriculture.com/news/livestock/colony-collapse-toll-is-highest-in-four-years-for-us-honeybees

Bee America. (2019, March 2). *What Is Honey, Exactly?* Retrieved 2021, from https://www.bee-america.com/content/what-honey-exactly

Randall, B. (2020, June 22). *The Value of Birds and Bees.* Farmers.Gov U.S. Department Of Agriculture. Retrieved 2021, from https://www.farmers.gov/connect/blog/conservation/value-birds-and-bees#:~:text=The%20plight%20of%20pollinators,types%20of%20fruits%20and%20vegetables

Riddle, S. (2016, July 25). The Chemistry of Honey. *Bee Culture: The Magazine of American Beekeeping.* https://www.beeculture.com/the-chemistry-of-honey/

U.S. Food and Drug Administration. (n.d.). *Helping Agriculture's Helpful Honey Bees: Drugs to Control American Foulbrood.* Retrieved 2021, from https://www.fda.gov/animal-veterinary/animal-health-literacy/helping-agricultures-helpful-honey-bees

Cockroach

Bohn, H., Nehring, V., G., J. R., & Klass, K. D. (2021). Revision of the genus Attaphila (Blattodea: Blaberoidea), myrmecophiles living in the mushroom gardens of leaf-cutting ants. *Arthropod Systematics & Phylogeny, 79*, 205–280. https://arthropod-systematics.arphahub.com/article/67569/list/9/

Heathcote, A. (2018, May 21). *Australia's giant burrowing cockroaches actually hiss.* Australian Geographic. Retrieved 2021, from https://www.australiangeographic.com.au/topics/wildlife/2018/05/australias-giant-burrowing-cockroaches-actually-hiss/

Mongeau, J.-M., McRae, B., Jusufi, A., Birkmeyer, P., Hoover, A. M., Fearing, R., & Full, R. J. (2012). Rapid Inversion: Running Animals and Robots Swing like a Pendulum under Ledges. *PLoS ONE.* Published.

https://journals.plos.org/plosone/article?id=10.1371/journal.pone.0038003

National Geographic. (n.d.). *Madagascar Hissing Cockroach.* Retrieved 2021, from https://www.nationalgeographic.com/animals/invertebrates/facts/madagascar-hissing-cockroach?loggedin=true

Quinn, M. (2019, February 26). *Species Attaphila fungicola - Ant Cockroach.* BugGuide. Retrieved 2021, from https://bugguide.net/node/view/1638767

Radford, B. (2010, August 9). *Could Cockroaches Really Survive A Nuclear Winter?* LiveScience. Retrieved 2021, from https://www.livescience.com/32749-could-cockroaches-really-survive-a-nuclear-winter.html

Ramel, G. (n.d.). *Blattodea 101: Your Guide To The Humble, Misunderstood Cockroach.* Earth Life. Retrieved 2021, from https://www.earthlife.net/insects/blatodea.html

Vanderbilt University. (2007, September 28). *Cockroaches Are Morons In The Morning, Geniuses In The Evening.* ScienceDaily. Retrieved 2021, from https://www.sciencedaily.com/releases/2007/09/070927132543.htm

Dolphin

Lewis, J. (2003). Mud Plume Feeding, a Unique Foraging Behavior of Bottlenose Dolphin in the Florida Keys. *Gulf of Mexico Science, 21*(1), 92–97. https://www.researchgate.net/publication/260338052_Mud_Plume_Feeding_a_Unique_Foraging_Behavior_of_the_Bottlenose_Dolphin_in_the_Florida_Keys

Sands, C. (n.d.). *When the Show's Over.* Dolphin Project. Retrieved 2021, from https://www.dolphinproject.com/blog/when-the-shows-over/

Smithsonian National Zoological Park. (2009, February 18). *Bottlenose Dolphins of Sarasota Bay.* Retrieved 2021, from https://web.archive.org/web/20090218004321/http:/nationalzoo.si.edu/ConservationAndScience/AquaticEcosystems/Dolphins/AboutDolphins/Reproduction.cfm

Whale and Dolphin Conservation. (n.d.). *Amazon River Dolphin.* Retrieved 2021, from https://us.whales.org/whales-dolphins/species-guide/amazon-river-dolphin/

Whale and Dolphin Conservation. (n.d.). *Whale and Dolphin Species Guide.* Retrieved 2021, from https://us.whales.org/whales-dolphins/species-guide/?gclid=EAIaIQobChMI1Z6s6IfC8QIV8PvjBx0n5gA5EAAYAiAAEgKX7PD_BwE

Willems, D. (2019). *Amazon River Dolphins.* World Wildlife Fund. https://wwfint.awsassets.panda.org/downloads/amazon_river_dolphin_2019.pdf

Woods Hole Oceanographic Institution. (n.d.). *How does a dolphin echolocate?* Retrieved 2021, from https://www.whoi.edu/science/B/people/kamaral/echolocation.html#:~:text=Dolphins%20and%20other%20toothed%20whales,the%20whale%20on%20food%20sources

Electric Eel

Smithsonian's National Zoo and Conservation Biology Institute. (n.d.). *Electric Eel.* Retrieved 2021, from https://nationalzoo.si.edu/animals/electric-eel

Echidna

A-Z Animals. (2021, September 26). *Echidna Tachyglossus Aculeatus.* Retrieved 2021, from https://a-z-animals.com/animals/echidna/

Bradford, A. (2016, December 19). *Facts About Echidnas.* LiveScience. Retrieved 2021, from https://www.livescience.com/57267-echidna-facts.html

Currumbin Wildlife Sanctuary. (2019, July 10). *10 Facts About Echidnas.* Retrieved 2021, from https://currumbinsanctuary.com.au/our-stories/news/10FactsaboutEchidnas

Government of South Australia Department for Environment and Water. (2020, October 9). *7 things you might not know about echidnas.* https://www.environment.sa.gov.au/goodliving/posts/2019/01/echidna-facts

Wong, E. S. W. (2013). Echidna Venom Gland Transcriptome Provides Insights into the Evolution of Monotreme Venom. *PLoS ONE*, *8*(11). https://journals.plos.org/plosone/article?id=10.1371/journal.pone.0079092

Frog

Ballance, A. (2009, February 19). *Frozen Frogs*. Radio New Zealand. Retrieved 2021, from https://www.rnz.co.nz/national/programmes/ourchangingworld/audio/1868835/frozen-frogs

Costanzo, J. P., do Amaral, M. C. F., Rosendale, A. J., & Lee, R. F., Jr. (2013). Hibernation physiology, freezing adaptation and extreme freeze tolerance in a northern population of the wood frog. *Journal of Experimental Biology*, *216*(18), 3461–3473. https://journals.biologists.com/jeb/article/216/18/3461/11609/Hibernation-physiology-freezing-adaptation-and

Foundation for National Parks and Wildlife. (n.d.). *Water holding frogs*. Backyard Buddies. Retrieved 2021, from https://backyardbuddies.org.au/backyard-buddies/water-holding-frogs/

LiveScience. (2012, September 26). *Can Frogs Survive Being Frozen?* https://www.livescience.com/32175-can-frogs-survive-being-frozen.html#:~:text=Fortunately%20for%20them%2C%20they%20don,and%20the%20western%20chorus%20frog

Mabin, S. (2020, March 21). *Desert frogs resurface after months — and sometimes years — underground waiting for rain*. ABC News Australia. Retrieved 2021, from https://www.abc.net.au/news/rural/2020-03-22/desert-frogs-resurface-after-rain/12071036

Naitoh, T., Wassersug, R., & Leslie, R., R. (1989). The Physiology, Morphology, and Ontogeny of Emetic Behavior in Anuran Amphibians. *Physiological Zoology*, *62*(3), 819–843. http://www.jstor.org/stable/30157929

THE BONKERS BIOLOGIST

Galápagos Tortoise
National Geographic. (n.d.-a). *Galápagos tortoise.* Retrieved 2021, from https://www.nationalgeographic.com/animals/reptiles/facts/galapagos-tortoise
National Geographic Kids. (n.d.). *Galápagos Tortoise.* Retrieved 2021, from https://kids.nationalgeographic.com/animals/reptiles/facts/galapagos-tortoise
San Diego Zoo Wildlife Alliance. (n.d.). *Galápagos Tortoise Chelonoidis nigra.* Retrieved 2021, from https://animals.sandiegozoo.org/animals/galapagos-tortoise

Greenland Shark
A-Z Animals. (2021, October 5). *Greenland Shark Somniosus microcephalus.* Retrieved 2021, from https://a-z-animals.com/animals/greenland-shark/
Lynnerup, N., Kjeldsen, H., Heegaard, S., Jacobsen, C., & Heinemeier, J. (2008). Radiocarbon Dating of the Human Eye Lens Crystallines Reveal Proteins without Carbon Turnover throughout Life. *PLoS ONE, 3*(1). https://journals.plos.org/plosone/article?id=10.1371/journal.pone.0001529
McGinty, J. C. (2020, September 11). How Old Is the Greenland Shark? The Answer Is Slippery. *Wall Street Journal.* https://www.wsj.com/articles/how-old-is-the-greenland-shark-the-answer-is-slippery-11599816602
O'Connor, M. R. (2017, November 25). The Strange and Gruesome Story of the Greenland Shark, the Longest-Living Vertebrate on Earth. *The New Yorker.* https://www.newyorker.com/tech/annals-of-technology/the-strange-and-gruesome-story-of-the-greenland-shark-the-longest-living-vertebrate-on-earth
Weisberger, M. (2017, December 14). *No, Scientists Haven't Found a 512-Year-Old Greenland Shark.* LiveScience. Retrieved 2021, from https://www.livescience.com/61210-shark-not-512-years-old.html

A-Z ANIMAL FACTS FOR KIDS

Hydra

Pappas, S. (2015, December 22). *Hail the Hydra, an Animal That May Be Immortal.* LiveScience. Retrieved 2021, from https://www.livescience.com/53178-hydra-may-live-forever.html

Rigby, S. (2019, June 3). Secret of the 'immortal' hydra's regenerating ability uncovered. *Science Focus.* https://www.sciencefocus.com/news/secret-of-the-immortal-hydras-regenerating-ability-uncovered/

Siebert, S., Farrell, J. A., Cazet, J. F., Abeykoon, Y., Primack, A. S., Schnitzler, C. E., & Juliano, C. E. (2019). Stem cell differentiation trajectories in Hydra resolved at single-cell resolution. *Science, 365*(6451). https://www.science.org/doi/10.1126/science.aav9314

Horseshoe Crab

A-Z Animals. (2021, June 15). *Horseshoe Crab.* Retrieved 2021, from https://a-z-animals.com/animals/horseshoe-crab/

Dais, M. (2019, March 18). *Horseshoe crabs are drained for their blue blood. That practice will soon be over.* Big Think. Retrieved 2021, from https://bigthink.com/health/horseshoe-crab-blue-blood/

Gatenby, C. (2016, November 3). *Modeling a Future for Horseshoe Crabs and Red Knots.* U.S. Fish and Wildlife Service. Retrieved 2021, from https://www.fws.gov/news/blog/index.cfm/2016/11/30/Modeling-a-Future-for-Horseshoe-Crabs-and-Red-Knots#:~:text=The%20red%20knot%2C%20a%20migratory,breeding%20grounds%20in%20the%20Arctic.&text=So%20much%20demand%20puts%20tremendous,crab%20population%20and%20red%20knots

U.S. Fish & Wildlife Service. (2006, August). *The Horseshoe Crab, Limulus polyphemus, A Living Fossil.* https://www.fws.gov/northeast/pdf/horseshoe.fs.pdf

Indian Giant Squirrel

A-Z Animals. (2021, October 11). *Indian Giant Squirrel Ratufa indica.* Retrieved 2021, from https://a-z-animals.com/animals/indian-giant-squirrel/

Bradford, A. (2014, June 27). *Squirrels: Diet, Habits & Other Facts.* LiveScience. Retrieved 2021, from https://www.livescience.com/28182-squirrels.html

Crew, B. (2014, October 9). *Giant black squirrels one of the longest animals.* Australian Geographic. Retrieved 2021, from https://www.australiangeographic.com.au/blogs/creatura-blog/2014/10/giant-black-squirrels-one-of-the-longest-animals/

Global Wildlife Conservation. (2019, January 21). *The Superheroes Of The Squirrel World: Flying Squirrels.* Re:Wild. Retrieved 2021, from https://www.rewild.org/news/the-superheroes-of-the-squirrel-world-flying-squirrels

Lawniczak, M. K. (n.d.). *Eastern grey squirrel Sciurus carolinensis.* BioKIDS. Retrieved 2021, from http://www.biokids.umich.edu/critters/Sciurus_carolinensis/

Sain, T., Sr. (n.d.). *Laotian Giant Flying Squirrel.* Our Breathing Planet. Retrieved 2021, from https://www.ourbreathingplanet.com/laotian-giant-flying-squirrel/

Solly, M. (2019, April 5). Yes, Giant Technicolor Squirrels Actually Roam the Forests of Southern India. *Smithsonian Magazine.* https://www.smithsonianmag.com/smart-news/yes-giant-technicolor-squirrels-actually-roam-forests-southern-india-180971886/

Indian Palm Squirrel

A-Z Animals. (2021, November 2). *Indian Palm Squirrel Funambulus Palmarum.* Retrieved 2021, from https://a-z-animals.com/animals/indian-palm-squirrel/

Government of Western Australia. (2018, May 2). *Northern palm squirrel.* Retrieved 2021, from https://www.agric.wa.gov.au/pest-mammals/northern-palm-squirrel

Jerboa

Colnect. (n.d.). *500 Tögrög (Wildlife Protection - Long-eared Jerboa).* Retrieved 2021, from https://colnect.com/en/coins/coin/25055-500_T%C3%B6gr%C3%B6g_Wildlife_Protection_-_Long-eared_Jerboa-2006~Today_-_Numismatic_Products_Wildlife_Protection-Mongolia

Hanney, P. W. (1975). *Rodents - their lives and habits*. David & Charles.

Lindstrom, H. (n.d.). *The Smallest Rodent*. Mongabay. Retrieved 2021, from https://rainforests.mongabay.com/kids/animals/smallest/smallest-rodent.html

Naish, D. (2017, October 3). *For Rodent Week, the Gift of Jerboas*. Scientific American. Retrieved 2021, from https://blogs.scientificamerican.com/tetrapod-zoology/for-rodent-week-the-gift-of-jerboas/

Power Coin. (n.d.). *LONG EARED JERBOA Gobi Desert Silver Coin 500 Togrog Mongolia 2006*. Retrieved 2021, from https://www.powercoin.it/en/cit-coin-invest-ag/2993-long-eared-jerboa-gobi-desert-silver-coin-500-togrog-mongolia-2006.html

Wikipedia. (n.d.). *Great Jerboa*. Retrieved 2021, from https://en.wikipedia.org/wiki/Great_jerboa

Wood, C. (2019, November 1). *What Is Convergent Evolution?* LiveScience. Retrieved 2021, from https://www.livescience.com/convergent-evolution.html

World Wildlife Fund. (2018). The long-eared jerboa stands—and hops—in a class of its own. *World Wildlife Magazine*. https://www.worldwildlife.org/magazine/issues/summer-2018/articles/the-long-eared-jerboa-stands-and-hops-in-a-class-of-its-own

Kookaburra

A-Z Animals. (2021c, June 27). *Kookaburra*. Retrieved 2021, from https://a-z-animals.com/animals/kookaburra/

Mayntz, M. (2019, April 2). *Do All Birds Migrate? No! Why Some Birds Stay Put All Year*. The Spruce. Retrieved 2021, from https://www.thespruce.com/why-birds-dont-migrate-4151247

THE BONKERS BIOLOGIST

Loggerhead Shrike

Institute for Wildlife Studies. (n.d.). *San Clemente Loggerhead Shrike Lanius ludovicianus mearnsi.* Retrieved 2021, from
https://www.iws.org/birds/san-clemente-loggerhead-shrike

San Diego Zoo Wildlife Alliance. (n.d.). *How We're Helping to Save the San Clemente Loggerhead Shrike.* Retrieved 2021, from
https://science.sandiegozoo.org/species/san-clemente-loggerhead-shrike

Lobster

DeCosta-Klipa, N. (2017, September 13). *Orange, yellow, blue, and even 'Halloween': The rarest lobster colors, explained.* Boston.Com. Retrieved 2021, from
https://www.boston.com/news/animals/2017/09/13/the-rarest-lobster-colors-explained/?amp=1

National Geographic. (n.d.). *The Weird World of Lobster Sex.* Retrieved 2021, from
https://www.nationalgeographic.com/magazine/article/basic-instincts-lobster-sex

The Lobster Conservancy. (2004). *Lobster Biology.* Retrieved 2021, from
http://www.lobsters.org/tlcbio/biology.html

The University of Maine Lobster Institute. (n.d.). *Anatomy & Biology.* Retrieved 2021, from
https://umaine.edu/lobsterinstitute/educational-resources/anatomy-biology/

Mantis Shrimp

Debczak, M. (2016, September 22). *10 Eye-Popping Facts About Mantis Shrimp.* Mental Floss. Retrieved 2021, from
https://www.mentalfloss.com/article/86128/10-eye-popping-facts-about-mantis-shrimp

What-When-How. (n.d.). *Stomatopoda (Mantis shrimps).* Retrieved 2021, from http://what-when-how.com/animal-life/order-stomatopoda/

Yong, E. (2008, July 18). *The Mantis Shrimp Has the World's Fastest Punch.* National Geographic. Retrieved 2021, from https://www.nationalgeographic.com/science/article/the-mantis-shrimp-has-the-worlds-fastest-punch

Nudibranch

National Geographic. (2009, December 2). *Nudibranchs*. Retrieved 2021, from https://www.nationalgeographic.com/animals/invertebrates/facts/nudibranchs-1

Oceana. (n.d.). *Blue Glaucus Glaucus atlanticus*. Retrieved 2021, from https://oceana.org/marine-life/corals-and-other-invertebrates/blue-glaucus

Silen, A. (n.d.). *Nudibranch*. National Geographic Kids. Retrieved 2021, from https://kids.nationalgeographic.com/animals/invertebrates/facts/nudibranch

Opossum

A-Z Animals. (2021, February 16). *Opossum Didelphis Virginiana*. Retrieved 2021, from https://a-z-animals.com/animals/opossum/

Geggel, L. (2019, March 3). *Why Are There So Many Marsupials in Australia?* LiveScience. Retrieved 2021, from https://www.livescience.com/64897-why-marsupials-in-australia.html

Kratt, C. (Writer), Kratt, M. (Writer), & Kratt, C. (Director), Kratt, M. (Director). (2014, July 9). Opossum in My Pocket (Season 3, Episode 8) [TV series episode]. In C. Kratt, M. Kratt, V. Commiso, B. Tohana (Executive Producers), *Wild Kratts*. Kratt Brothers Company and 9 Story Media Group.

National Opossum Society. (n.d.). *The National Opossum Society welcomes you to the world of the Virginia opossum!* Retrieved 2021, from https://www.opossum.org/

Oyster

Beck, M. W., Brumbaugh, R. D., Airoldi, L., Carranza, A., Coen, L. D., Crawford, C., Defeo, O., Edgar, G. J., Hancock, B., Kay, M., Lenihan,

H., Luckenbach, M. W., Toropova, C. L., & Zhang, G. (2009). *Shellfish Reefs at Risk: A Global Analysis of Problems and Solutions.* The Nature Conservancy. https://www.conservationgateway.org/ConservationPractices/Marine/Documents/Shellfish%20Reefs%20at%20Risk-06.18.09-Pages.pdf
Chesapeake Bay Foundation. (n.d.). *Oyster Fact Sheet.* Retrieved 2021, from https://www.cbf.org/about-the-bay/more-than-just-the-bay/chesapeake-wildlife/eastern-oysters/oyster-fact-sheet.html
Chesapeake Bay Program. (n.d.). *Oysters.* Retrieved 2021, from https://www.chesapeakebay.net/issues/oysters
Horn Point Laboratory Oyster Hatchery. (2021). *Our Facilities.* University of Maryland Center for Environmental Science Horn Point Laboratory Oyster Hatchery. Retrieved 2021, from https://hatchery.hpl.umces.edu/facility-descriptions/
United States Environmental Protection Agency. (n.d.). *Deepwater Horizon – BP Gulf of Mexico Oil Spill.* Retrieved 2021, from https://www.epa.gov/enforcement/deepwater-horizon-bp-gulf-mexico-oil-spill#:~:text=On%20April%2020%2C%202010%2C%20the,of%20marine%20oil%20drilling%20operations

Platypus

Australian Platypus Conservancy. (2020, November). *Platypus News & Views* (No. 82). https://platypus.asn.au/wp-content/uploads/2020/11/e-pnv-82.pdf
Australian Platypus Conservancy. (n.d.). *Some FAQs.* Retrieved 2021, from https://platypus.asn.au/faqs/
Burgin, C. J., Colella, J. P., Kahn, P. L., & Upham, N. S. (2018). How many species of mammals are there? *Journal of Mammology, 99*(1), 1–14. https://doi.org/10.1093/jmammal/gyx147
National Oceanic and Atmospheric Administration. (n.d.). *What is a platypus?* National Ocean Service. Retrieved 2021, from https://oceanservice.noaa.gov/facts/platypus.html

A-Z ANIMAL FACTS FOR KIDS

Paddlefish

A-Z Animals. (2021, February 17). *Paddlefish*. Retrieved 2021, from https://a-z-animals.com/animals/paddlefish/

Montana Natural Heritage Program and Montana Fish, Wildlife and Parks. (n.d.). *Montana Field Guides: Paddlefish - Polyodon spathula*. Montana's Official State Website. Retrieved 2021, from http://fieldguide.mt.gov/speciesDetail.aspx?elcode=AFCAB01010#:~:text=Although%20it%20is%20sometimes%20called,Montana%20AFS%20Species%20Status%20Account)

U.S. National Park Service. (n.d.). *Invasive Zebra Mussels*. Retrieved 2021, from https://www.nps.gov/articles/zebra-mussels.htm

Queensland Grouper

Aquarium of the Pacific. (n.d.). *Queensland Grouper Epinephelus lanceolatus*. Retrieved 2021, from https://www.aquariumofpacific.org/onlinelearningcenter/species/queensland_grouper

Marine Education Society of Australasia. (2015). *Queensland giant grouper*. Retrieved 2021, from http://www.mesa.edu.au/atoz/giant_grouper.asp

Ruby-Throated Hummingbird

A-Z Animals. (2021, July 4). *Ruby-Throated Hummingbird Archilochus colubris*. Retrieved 2021, from https://a-z-animals.com/animals/ruby-throated-hummingbird/

BirdNote. (2017, August 21). *Why Only Certain Birds Can Drink Nectar: The sugary food source might give some species an adaptive advantage*. Audubon. Retrieved 2021, from https://www.audubon.org/news/why-only-certain-birds-can-drink-nectar

Hummingbird Central. (2021). *Hummingbird Facts and Family Introduction*. Retrieved 2021, from https://www.hummingbirdcentral.com/hummingbird-facts.htm

Hummingbird-Guide.com. (2021). *Hummingbird Banding: A Key Tool for Their Survival*. Retrieved 2021, from https://www.hummingbird-guide.com/hummingbird-banding.html

National Geographic. (2020, June 15). *Hummingbirds see colors we can't even imagine*. Retrieved 2021, from https://www.nationalgeographic.com/animals/article/hummingbirds-see-colors-outside-rainbow?loggedin=true

Shewey, J. (2021, April 7). *How Do Hummingbirds Use Their Tongues and Beaks?* Birds & Blooms. Retrieved 2021, from https://www.birdsandblooms.com/birding/attracting-hummingbirds/hummingbird-tongues-beaks/

Stoddard, M. C., Eyster, H. N., Hogan, B. G., Morris, D. H., Soucy, E. R., & Inouye, D. W. (2020). Wild hummingbirds discriminate nonspectral colors. *Proceedings of the National Academy of Sciences, 117*(26), 15112–15122. https://www.pnas.org/content/117/26/15112

Sea Turtle

A-Z Animals. (2021, September 24). *Sea Turtle*. Retrieved 2021, from https://a-z-animals.com/animals/sea-turtle/

National Oceanic and Atmospheric Administration. (n.d.-a). *Sea Turtles: What do you know about one of the world's most endangered species?* National Ocean Service. Retrieved 2021, from https://oceanservice.noaa.gov/news/june15/sea-turtles.html#:~:text=It%20takes%2020%2D30%20years,until%20the%20age%20of%2080

Poovey, C. C. (2014, July 10). Karen Beasley's legacy: save the turtles. *Wake Forest Magazine*. https://magazine.wfu.edu/2014/07/10/karen-beasleys-legacy-save-the-turtles/

Skwarecki, B. (2014, June 1). Where Do Baby Sea Turtles Go? Tracking the transatlantic journey of young sea turtles reveals surprises. *Scientific American, 310*(6). https://www.scientificamerican.com/article/where-do-baby-sea-turtles-go/#:~:text=After%20baby%20loggerhead%20turtles%20hatch,years%20near%20those%20same%20beaches

A-Z ANIMAL FACTS FOR KIDS

World Wildlife Fund. (n.d.). *How Long Do Sea Turtles Live? And Other Sea Turtle Facts.* Retrieved 2021, from https://www.worldwildlife.org/stories/how-long-do-sea-turtles-live-and-other-sea-turtle-facts

World Wildlife Fund. (n.d.-a). *Infographic: Sea Turtles.* Retrieved 2021, from https://www.worldwildlife.org/pages/infographic-sea-turtles

Termite

Department of Systematic Biology, Entomology Section, National Museum of Natural History. (1996). *Incredible Insects.* Smithsonian Institution. Retrieved 2021, from https://www.si.edu/spotlight/buginfo/incredbugs#:~:text=The%20Longest%2Dlived%20Insect%3A%20The,they%20live%20for%20100%20years

Martius, C., & d'Arc Ribeiro, J. (1996). Colony Populations and Biomass in Nests of the Amazonian Forest Termite Anoplotermes banksi Emerson (Isoptera: Termitidae). *Studies on Neotropical Fauna and Environment, 31*(2), 82–86. https://www.tandfonline.com/doi/abs/10.1076/snfe.31.2.82.13328?journalCode=nnfe20&#:~:text=Colony%20size%20of%207%20nests,Queens%20weighed%2010%E2%80%9330%20mg

Hadley, D. (2019, February 16). *10 Fascinating Facts About Termites.* ThoughtCo. Retrieved 2021, from https://www.thoughtco.com/fascinating-facts-about-termites-1968587

Orkin. (n.d.). *What Do Soldier Termites Do In A Termite Colony?* Retrieved 2021, from https://www.orkin.com/termites/colony/soldier-termite

Tamandua

A-Z Animals. (2021, April 17). *Anteater Myrmecophaga Tridactyla.* Retrieved 2021, from https://a-z-animals.com/animals/anteater/

San Diego Zoo Wildlife Alliance. (n.d.-a). *amandua or Lesser Anteater, Tamandua tetradactyla, T. mexicana.* Retrieved 2021, from https://animals.sandiegozoo.org/animals/tamandua-or-lesser-anteater

Wilson, J. (2010, December 2). *New Anteater Species, Tamandua!* Sacramento Zoo. Retrieved 2021, from https://www.saczoo.org/2010/12/new-anteater-species-tamandua/#:~:text=Like%20the%20giant%20anteater%2C%20the,each%20day%20in%20the%20wild.&text=A%20smelly%20scent%20gives%20tamanduas,pungent%20than%20a%20skunk's%20scent

Wood, W. F. (1999). The History of Skunk Defensive Secretion Research. *The Chemical Educator, 4*(2), 44–50. https://web.archive.org/web/20030902091556/http://chemeducator.org/sbibs/s0004002/spapers/420044ww.pdf

Uakari

A-Z Animals. (2021, February 16). *Uakari Cacajao.* Retrieved 2021, from https://a-z-animals.com/animals/uakari/

Groves, C. P. (2004, April 22). *Monkey Primate.* Britannica. Retrieved 2021, from https://www.britannica.com/animal/monkey

Unau

Delsuc, F., Kuch, M., Gibb, G. C., Karpinski, E., Hackenberger, D., Szpak, P., Martínez, J. G., Mead, J. I., McDonald, H. G., MacPhee, R. D. E., Billet, G., Hautier, L., & Poinar, H. N. (2019). Ancient Mitogenomes Reveal the Evolutionary History and Biogeography of Sloths. *Current Biology, 29*(12), 2031–2042. https://www.sciencedirect.com/science/article/pii/S0960982219306135X

Presslee, S., Slater, G. J., Pujos, F., Forasiepi, A. M., Fischer, R., Molloy, K., Mackie, M., Olsen, J. V., Kramarz, A., Taglioretti, M., Scaglia, F., Lezcano, M., Lanata, J. L., Southon, J., Feranec, R., Bloch, J., Hajduk, A., Martin, F. M., Gismondi, R. S., . . . MacPhee, R. D. E. (2019). Palaeoproteomics resolves sloth relationships. *Nature Ecology & Evolution, 3,* 1121–1130. https://doi.org/10.1038/s41559-019-0909-z

Science News Staff. (2019, June 7). DNA and Protein Studies Shake Up Sloth Family Tree. *Science News.* http://www.sci-news.com/biology/sloth-family-tree-07268.html

Sloth Conservation Foundation. (2016). *10 incredible facts about the sloth.* Retrieved 2021, from https://slothconservation.org/10-incredible-facts-about-the-sloth/?gclid=CjwKCAjw95yJBhAgEiwAmRrutPQ6IR_ivDxYM7_0tqxsSSFVx07g4xNvk9s8yk0WuNuAUQao2YBq3xoCIJMQAvD_BwE

Sloth Conservation Foundation. (2016b). *With a little help from my friends: sloth hair, moths and algae.* Retrieved 2021, from https://slothconservation.org/with-a-little-help-from-my-friends-sloths-moths-and-algae/#:~:text=These%20moths%20are%20exclusively%20found,a%20three%2Dfingered%20sloth's%20fur

Vogel, G. (2019, June 6). Ancient molecules reveal surprising details on origins of 'bizarre' sloths: Analysis of fossil molecules suggest creatures may have journeyed across ancient land bridge. *Science.* https://www.science.org/content/article/ancient-molecules-reveal-surprising-details-origins-bizarre-sloths

World Wildlife Fund. (n.d.-c). *Sloth Facts.* Retrieved 2021, from https://www.worldwildlife.org/species/sloth#:~:text=There%20are%20two%20different%20types,%2Dthroated%20sloth%20(Bradypus%20tridactylus)

Vampire Squid

A-Z Animals. (2021, February 24). *Vampire Squid Vampyroteuthis infernalis.* Retrieved 2021, from https://a-z-animals.com/animals/vampire-squid/

Aquarium of the Pacific. (2021). *Vampire Squid Vampyroteuthis infernalis.* Retrieved 2021, from https://www.aquariumofpacific.org/onlinelearningcenter/species/vampire_squid

Johnson, B. (2000). *Vampyroteuthis infernalis.* Animal Diversity Web. Retrieved 2021, from https://animaldiversity.org/accounts/Vampyroteuthis_infernalis/

THE BONKERS BIOLOGIST

MarineBio. (2021). *Vampire Squid, Vampyroteuthis infernalis.* Retrieved 2021, from https://www.marinebio.org/species/vampire-squid/vampyroteuthis-infernalis/

Oceana. (n.d.-b). *Vampire Squid Vampyroteuthis infernalis.* Retrieved 2021, from https://oceana.org/marine-life/cephalopods-crustaceans-other-shellfish/vampire-squid

Virgin Islands Dwarf Gecko

A-Z Animals. (2021, July 11). *Virgin Islands Dwarf Gecko Sphaerodactylus parthenopion.* Retrieved 2021, from https://a-z-animals.com/animals/virgin-islands-dwarf-gecko/

Dickerson, K. (2014, August 12). *Geckos' Sticky Secret? They Hang by Toe Hairs.* LiveScience. Retrieved 2021, from https://www.livescience.com/47307-how-geckos-stick-and-unstick-feet.html

Fox, A. (2021, February 3). Chameleon Discovered in Madagascar May Be World's Smallest Reptile: The male of the newly described species measured just half an inch long from his nose to the base of his tail. *Smithsonian Magazine.* https://www.smithsonianmag.com/smart-news/chameleon-discovered-madagascar-may-be-worlds-smallest-reptile-180976909/

Guinness World Records. (2021). *Smallest living gecko species.* Retrieved 2021, from https://www.guinnessworldrecords.com/world-records/596062-smallest-living-gecko-species

Leclair, R., Jr, & Leclair, M. H. (2011). Life-History Traits in a Population of the Dwarf Gecko, Sphaerodactylus vincenti ronaldi, from a Xerophytic Habitat in Martinique, West Indies. *Copeia, 2011*(4), 566–576. https://www.jstor.org/stable/41416577

National Geographic. (2021, February 1). *New chameleon species may be world's smallest reptile.* Retrieved 2021, from https://www.nationalgeographic.com/animals/article/tiny-chameleon-smallest-reptile-discovered-madagascar?loggedin=true

Whelk

Diep, F. (2014, January 14). Six Years After Chemical Ban, Fewer Female Snails Are Growing Penises: Before the ban, in the worst cases, the added male plumbing would make female snails explode. *Popular Science.* https://www.popsci.com/article/science/six-years-after-chemical-ban-fewer-female-snails-are-growing-penises/

Draxler, B. (2014, January 15). Female Sea Snails No Longer Growing Penises Thanks to Ban on Toxic Chemical. *Discover Magazine.* https://www.discovermagazine.com/environment/female-sea-snails-no-longer-growing-penises-thanks-to-ban-on-toxic-chemical

Goldstein, M. (2010, September 15). *Female snails in Australia are just happy to see you.* Deep Sea News. Retrieved 2021, from http://www.deepseanews.com/2010/09/female-snails-in-australia-are-just-happy-to-see-you/

Purcell, C. (2004, October 21). *Toxic paint makes females grow penis.* ABC Science. Retrieved 2021, from https://www.abc.net.au/science/articles/2004/10/21/1224721.htm#:~:text=Almost%20half%20of%20female%20sea,algae%20growing%20on%20boat%20hulls

Whale Shark

Hart, T., & Buzzacott, P. (n.d.). *Diver Was Half-Swallowed by a Whale Shark.* Divers Alert Network. Retrieved 2021, from https://dan.org/safety-prevention/diver-safety/case-summaries/diver-was-virtually-swallowed-by-a-whale-shark/#:~:text=The%20diver%20recalls%20being%20hit,her%20about%20in%20the%20water

Ocean Conservancy. (2021). *Wildlife Fact Sheets: Whale Shark RHINCODON TYPUS.* Retrieved 2021, from https://oceanconservancy.org/wildlife-factsheet/whale-shark/

Xingu River Ray

Lincoln Park Zoo. (n.d.). *White-blotched River Stingray.* Retrieved 2021, from https://www.lpzoo.org/animal/white-blotched-river-stingray/

National Aquarium. (2021). *White-Blotched River Stingray (Potamotrygon leopoldi)*. Retrieved 2021, from
   https://aqua.org/explore/animals/white-blotched-river-stingray

National Geographic Society. (2021). *Xingu River Ray*. Retrieved 2021, from
   https://www.nationalgeographic.org/projects/photo-ark/animal/potamotrygon-leopoldi/

Smithsonian's National Zoo and Conservation Biology Institute. (n.d.-b). *Freshwater Stingray*. Retrieved 2021, from
   https://nationalzoo.si.edu/animals/freshwater-stingray

Xanthippe's Shrew

Bryan, V. (2014, March 3). *Five Reasons Why Socrates Was A Terrible Husband*. Classical Wisdom. Retrieved 2021, from
   https://classicalwisdom.com/people/philosophers/five-reasons-socrates-terrible-husband/

Havahart. (n.d.). *Shrews*. Retrieved 2021, from
   https://www.havahart.com/shrew-facts

Our Endangered World. (2021, September 30). *15 Animals that Start with X*. Retrieved 2021, from
   https://www.ourendangeredworld.com/species/animals-that-start-with-x/#3_Xanthippes_Shrew

White-toothed shrew. (2021, September 17). Wikipedia. Retrieved 2021, from https://en.wikipedia.org/wiki/White-toothed_shrew

Yeti Crab

A-Z Animals. (2021, August 7). *Yeti Crab*. Retrieved 2021, from https://a-z-animals.com/animals/yeti-crab/

Cande, S. C., Raymond, C. A., Stock, J., & Haxby, W. F. (1995). Geophysics of the Pitman Fracture Zone and Pacific-Antarctic Plate Motions During the Cenozoic. *Science, 270*(5238), 947–953.
   https://www.science.org/doi/10.1126/science.270.5238.947

Macpherson, E., Jones, W., & Segonzac, M. (2005). A new squat lobster family of Galatheoidea (Crustacea, Decapoda, Anomura) from the hydrothermal vents of the Pacific-Antarctic Ridge. *Zoosystema, 27*(4), 709–723.

https://www.researchgate.net/publication/228652104_A_new_squat_lobster_family_of_Galatheoidea_Crustacea_Decapoda_Anomura_from_the_hydrothermal_vents_of_the_Pacific-Antarctic_Ridge

National Oceanic and Atmospheric Administration. (2018). *How does the temperature of ocean water vary?* Ocean Exploration. Retrieved 2021, from https://oceanexplorer.noaa.gov/facts/temp-vary.html

National Oceanographic and Atmospheric Administration. (2021, February 26). *What is a hydrothermal vent? Hydrothermal vents form at locations where seawater meets magma.* National Ocean Service. Retrieved 2021, from https://oceanservice.noaa.gov/facts/vents.html#:~:text=Seawater%20in%20hydrothermal%20vents%20may,where%20the%20vents%20are%20formed

Zebra Shark

A-Z Animals. (2021, February 15). *Zebra Shark Stegostoma Fasciatum.* Retrieved 2021, from https://a-z-animals.com/animals/zebra-shark/

National Geographic. (2020, August 25). *Parthenogenesis, explained: How some animals have "virgin births."* Retrieved 2021, from https://www.nationalgeographic.com/animals/article/parthenogenesis-how-animals-have-virgin-births

Zigzag Salamander

Amphibians and Reptiles of North Carolina. (2021). *Southern Zigzag Salamander Plethodon ventralis.* Retrieved 2021, from https://herpsofnc.org/southern-zigzag-salamander/

Conant, R., & Collins, J. T. (1998). *Reptiles and Amphibians of Eastern and Central North America* (Third Edition, Expanded ed.). Houghton Mifflin, New York, NY.

Davis, J. (2020, June 22). *Metamorphosis is helping to explain salamander skull diversity.* Natural History Museum. Retrieved 2021, from https://www.nhm.ac.uk/discover/news/2020/june/metamorphosis-is-helping-to-explain-salamander-skull-diversity.html

Indiana Herp Atlas. (n.d.). *Northern Zigzag Salamander.* Retrieved 2021, from https://www.inherpatlas.org/species/plethodon_dorsalis

Missouri Department of Conservation. (n.d.). *Ozark Zigzag Salamander*. Retrieved 2021, from https://mdc.mo.gov/discover-nature/field-guide/ozark-zigzag-salamander

Tennessee Wildlife Resources Agency. (n.d.). *Northern Zigzag Salamander Plethodon dorsalis*. Retrieved 2021, from https://www.tn.gov/twra/wildlife/amphibians/salamanders/northern-zigzag-salamander.html

Virginia Herpetological Society. (2021). *Southern Zigzag Salamander Plethodon ventralis*. Retrieved 2021, from https://www.virginiaherpetologicalsociety.com/amphibians/salamanders/southern-zigzag-salamander/southern_zigzag_salamander.php

Zug, G. R. (2008, May 2). *Lungless Salamander, Amphibian*. Britannica. Retrieved 2021, from https://www.britannica.com/animal/lungless-salamander

Printed in Great Britain
by Amazon

b60cc4d3-5878-4775-80aa-c5b0289c1c32R01